THE WORLD
OF TOUCH

David Katz

THE WORLD OF TOUCH

Edited and translated by

Lester E. Krueger

The Ohio State University

LAWRENCE ERLBAUM ASSOCIATES, PUBLISHERS
1989   Hillsdale, New Jersey   Hove and London

Licensed edition with permission of the original publisher, Johann Ambrosius Barth in Leipzig. Translated from *Der Aufbau der Tastwelt* (The world of touch). Ergaenzungsband 11 (Supplement Volume 11), *Zeitschrift fuer Psychologie und Physiologie der Sinnesorgane*, Abt. I (Part I), *Zeitschrift fuer Psychologie*. Copyright 1925 by Johann Ambrosius Barth, Leipzig.

Copyright © 1989 by Lawrence Erlbaum Associates, Inc.
All rights reserved. No part of this book may be reproduced in any form, by photostat, microfilm, retrieval system, or any other means, without the prior written permission of the publisher.

Lawrence Erlbaum Associates, Inc., Publishers
365 Broadway
Hillsdale, New Jersey 07642

Library of Congress Cataloging-in-Publication Data

Katz, David, 1884–1953.
    [Aufbau der Tastwelt. English]
    The world of touch / David Katz; edited and translated by Lester E. Krueger.
    p. cm.
    Translation of Der Aufbau der Tastwelt. Published as Ergaenzungsband 11, Zeitschrift fuer Psychologie und physiologie der Sinnesorgane, Abt. I, Zeitschrift fuer Psychologie.
    Bibliography: p.
    Includes indexes.
    ISBN 0-8058-0529-X
    1. Touch.   I. Krueger, Lester E.   II. Title.
BF275.K3713   1989
152.1'82—dc20   89-11868
                                            CIP

Printed in the United States of America
10 9 8 7 6 5 4 3 2 1

To my beloved wife, Rosa Katz

Contents

Translator's Preface   xi

Editor's Introduction   *1*

    Touch vs. Vision   *2*
    Touch Redux   *3*
    The Intelligent Hand   *4*
    Vibration Sense   *5*
    Texture   *8*
    Katz's Phenomenology   *12*
    References   *17*

Author's Preface   *23*

DIVISION I: MODES OF APPEARANCE OF THE WORLD OF TOUCH   *25*

    Section 1. Introduction   *27*

Editor's Notes   *31*

Chapter I: Critical and Methodological Preamble   *33*

    Section 2. The Atomistic Viewpoint of Sensory Psychology   *33*
    Section 3. The More Complex Tactual Phenomena in the Present Literature   *37*
    Section 4. Visual and Tactual Perception   *39*

Editor's Notes   *45*

Chapter II: Types of Tactual Phenomena   *47*

    Section 5. The Monotony of Tactual Matter and the Polymorphism of its Modes of Appearance   *47*
    Section 6. Surface Touch and Immersed Touch   *50*
    Section 7. Volume Touch   *52*
    Section 8. Touch-Transparent Film   *53*
    Section 9. Qualities (*Modifikationen*) and Identifying Characteristics (*Spezifikationen*) of Surface Touch   *54*
    Section 10. Natural and Artificial Forms of Materials   *56*
    Section 11. Continuity of the Tactual Surface; Tactual Figure and Ground   *59*
    Section 12. Tactual Images of Things: Memory Touches   *62*
    Addendum. Eidetic Tactual Images and Tactual Hallucinations   *68*
    Section 13. Refutation of Possible Objections   *69*

Editor's Notes   *73*

Chapter III : Movement as a Formative Factor in Tactual Phenomena   *75*

    Section 14. The Bias Toward Temporal Atomism in the Approach of Previous Studies on the Psychology of Perception   *75*
    Section 15. The Stationarity Principle in the Methodology of Earlier Tactual Experiments   *77*
    Section 16. Movement as a Creative Force in the Sense of Touch   *79*
    Addendum. Movement as a Creative Force in Other Sensory Domains   *82*
    Section 17. Constancy of the Position of Objects with a Moving Touch Organ   *84*
    Addendum. Kinetic and Motionless Figures for Humans and Animals   *86*

Editor's Notes   *89*

DIVISION II : QUANTITATIVE STUDIES OF THE TACTUAL PERFORMANCE   *91*

Chapter I : Studies on Surface Touch   *93*

Part I: Experiments on the Qualities (*Modifikationen*) of Surface Touch   *93*

    Section 18. The Tactual Material   *93*
    Section 19. Basic Experiment   *94*
    Section 20. Variation in the Size of the Tactual Surfaces   *96*
    Section 21. The Reduction of Tactual Impressions   *98*
    Section 22. Touching without Lateral Movements on the Tactual Surface   *100*
    Section 23. Experiments with Movement of the Tactual Surfaces   *101*

CONTENTS ix

       Section 24. Veiling Intermediaries   *109*
       Section 25. Remote Touching, Especially in Medical Practice   *121*
       Section 26. Tactual Performance of Body Parts not Usually Used for Touching   *123*
       Section 27. One's Own Body as a Tactual Object   *126*
       Section 28. Experiments with Adapted Touch Organs   *127*

Part II: Experiments of the Identifying Characteristics (*Spezifikationen*) of Surface Touch   *130*

       Section 29. Recognition Time for the Identifying Characteristics (*Spezifikationen*) of Surface Touch   *130*
       Section 30. Experiments with Amputees   *138*

Editor's Notes   *143*

Chapter II. Studies on Touch Transparency   *145*

       Section 31. Sensitivity to Thickness   *145*
       Section 32. Touch-Transparency of Space   *149*

Editor's Notes   *151*

Chapter III: Analysis of Touch Movements   *153*

       Section 33. Graphically Recording the Movement of a Single Finger   *153*
       Section 34. Movement of the Fingers in their Natural Configuration   *156*
       Section 35. Deformation of the Fingertip in Touching   *160*

Editor's Notes   *163*

Chapter IV: The Role of the Temperature Sensation in Touch   *165*

       Section 36. Thermal Qualities of Objects at Varied Temperatures   *165*
       Addendum. Perspiration and Touching   *174*
       Section 37. Temperature Gestalt and the Thermal Capacity of Objects   *175*
       Section 38. Fashioning of the Temperature Impressions   *177*

Editor's Notes   *183*

DIVISION III: FURTHER ANALYSIS OF THE TACTUAL PERFORMANCE   *185*

Chapter I: The Role of the Vibration Sense in Touch   *187*

       Section 39. Vibration Sensations in Neurological Research and in the Lives of Deaf-Mutes   *187*
       Section 40 Vibratory Thresholds: Rejection of Visual Analogies   *194*
       Section 41. The Independence of Vibration and Pressure Sensations   *197*

Section 42. The Form of Vibratory Stimulation, and the Temporal Pattern of the Vibration Sensation   *200*
Section 43. Perception of Vibration States through the Vibration Sense   *203*
Section 44. The Sensory Organs of the Vibration Sense   *209*
Section 45. Perception of Object Properties Through the Vibration Sense   *212*
Addendum. The Auditory Recognition of Object Properties   *221*

Editor's Notes   *223*

Chapter II: The Contribution of Visual and Kinesthetic Processes to the Structure of Tactual Forms   *225*

Section 46. The Influence of Visual Representations   *225*
Section 47. The Influence of Kinesthetic Processes   *232*

Editor's Notes   *235*

DIVISION IV: APPLICATIONS   *237*

Section 48. The Psychology of Language and Perception   *238*
Section 49. The Sense of Touch in Education   *246*
Section 50. The Sense of Touch in Psychological Testing   *247*

Author Index   *251*

Subject Index   *257*

Translator's Preface

The goal of this translation was to put David Katz's *Der Aufbau der Tastwelt* (The World of Touch) into as natural and correct a form of modern American English as possible. To accomplish that objective, Katz's long German sentences were sometimes divided into several English sentences, or somewhat condensed into more compact English form, as long as the intended meaning was thereby preserved. Materials inserted into the text for clarification, etc., are set off by brackets. More extended comments on the text are provided before each chapter or division, and in the Editor's Introduction. Katz's previous publications in English (e.g., Katz, 1930, 1935) provided useful guides on how to translate particular words and phrases. I deviated from Katz (1935), however, by generally translating *Materialstruktur* as "texture," rather than "material structure."

Several persons made notable contributions to the present work. Thomas G. R. Bower gave the initial impetus to the project when he suggested that I prepare an English summary of *Tastwelt* (Krueger, 1970, 1982) for his graduate seminar in perception at Harvard University. Dale S. Cunningham assisted in the preparation of a preliminary version of the translation, and he provided the English translations for the Greek and Latin terms and phrases that Katz used. Helpful comments were provided by the late Francis P. Hardesty on the preliminary version of the translation, and by Morton A. Heller and Leann Stadtlander on the penultimate version. Ms. Stadtlander also prepared the author and subject indexes; the latter was devised from scratch, since Katz provided no subject index in the original German edition. Susan Lederman provided very extensive (and very useful!) comments on an intermediate draft. Lisa A. Meyer and Jacquelyn A. Reilly showed great perseverance and efficiency in typing and revising the word-processor files on the project. David Katz (or his spirit) also provided assistance for the project by revealing some interesting facet of tactual perception on virtually every page of the volume. That made the task much less of a burden, and more of a journey of discovery.

Many of Katz's sources are inaccessible, and therefore his references generally were not modified or put into a more modern style. Katz's system of footnotes was used, except that the footnotes and references now are listed at the end of each section, rather than at the bottom of each page. In some cases, such as for Katz's citations of his own work, however, the year is given in parentheses after the name, and the full reference is provided at the end of the Editor's Introduction which follows next.

Lester E. Krueger

Editor's Introduction

David Katz (1884-1953) was a major figure in the study of the psychology of perception. His more than 100 publications include at least 20 substantial books and monographs (see obituaries by Arnheim, 1953, and MacLeod, 1954). His career began in Germany, shifted to England in 1933, and ended in Sweden, where he was Professor Emeritus at the University of Stockholm. It roughly spanned that of the Gestalt movement, and indeed in his sympathies Katz "stood closest to the Gestalt theorists" (MacLeod, 1954, p. 3). He even wrote a textbook on Gestalt psychology (Katz, 1950). Katz was never a true Gestalt psychologist, however. Texture and ground (microstructure) concerned him more than did form and figure (macrostructure), and on this and several other points, he was closer to the position of James J. Gibson (1966, 1979). Whereas the Gestaltists focused on the simplicity of the internal response, Katz, like Gibson, focused on its veridicality—that is, its correspondence with the external stimulus.

Like Gibson, too, Katz's work in perception mainly involved touch and vision. Two of his most important works are *Tastwelt* (World of Touch) and *Farbwelt* (World of Color). *Tastwelt* first appeared in German in 1925 and is presented here in English translation. English summaries or synopses of *Tastwelt* have been provided by Katz (1930), Krueger (1970), and Zigler (1926). A wider-ranging discussion of *Tastwelt* and related work by Katz and others was provided by Krueger (1982).

Other researchers also have been active in the area of tactual perception. Reference chapters or volumes on tactual research in general have been provided by Geldard (1974), Gordon (1978), Kennedy (1978), Loomis and Lederman (1987), and Schiff and Foulke (1982), and the interested reader also may wish to see the tactual studies by von Békésy (1967), Gibson (1962), Lederman and Klatzky (1987), and Révész (1950). Relatively little work has been done on the higher-order properties of touch since Katz's monograph appeared, however, so the present English translation of the Katz volume should be of considerable contemporary as well as historical interest.

Touch vs. Vision

Lederman (personal communication, June 1988) cited four reasons why our knowledge of touch has not developed as rapidly as that of vision and audition: 1) there is a lack of off-the-shelf technology available for producing/presenting tactual stimulus arrays to the skin, 2) monks during medieval times were not allowed to transcribe, and therefore could not preserve, any scientific writings pertaining to the skin, 3) there is a general reluctance in our society to discuss touch-related matters, and 4) the skin is the largest organ in the body, and highly complex.

In part, too, touch has received less attention because vision typically dominates or "captures" touch when the two modalities are in conflict (Rock & Harris, 1967; but see Section 11 of the World of Touch, and Heller, 1983). According to Révész (1950), "when we touch some common object, the tactile impression is always permeated with visual experiences" (p. 156), and seeing is indispensable to the sculptor. However, Katz wanted to dispel the invidious distinction between the supposedly higher (e.g., vision, audition) and lower (e.g., touch) senses. In a fascinating linguistic analysis (Section 48), he showed how touch and the hand have influenced language, providing, in particular, many terms for cognitive activities (e.g., grasp, handle, comprehend).

By showing how wondrous are the abilities of touch and how rich the tactual stimulus can be in specifying objects, surfaces, substances, and events, he hoped to regain for touch its former prominence, if not its predominance. In touching, Katz noted, one brings object properties to life, creating through one's own muscular activity such qualities as roughness and smoothness, and hardness and softness (Section 48). "The tactual properties of our surroundings do not chatter at us like their colors; they remain mute until we make them speak." Eyemovements may endow vision with a certain activity, he conceded, but "eyemovements do not create color the way finger movements create touch."

The fingers, as wielded by the hand, Katz noted, obtain information on the innards of objects, whereas the eye, remaining fixed at the outer surface of objects, plays a lesser role in developing the belief in the reality of the external world (Section 48). Such basic concepts of physics as force, impenetrability, resistance, and friction are rooted in touch. However, Katz provided no data of his own, but merely adopted the classical philosophical position which advocated the primacy of touch (Krueger, 1982). Also, he said that visual imagery permeates touch, and that vision was absolutely crucial in providing the quality of spatiality to the tactual representation (Section 46). As noted above, more recent data indicate that, in general, vision dominates and educates touch, not vice versa (Rock & Harris, 1967). Thus, vision plays an important role in guiding touch and in interpreting tactual behavior and experience.

Katz found that the fingers surpass the eye in judging the thickness of paper (Section 31) and in detecting vibrations (Section 41). According to him, the fingers are relatively more sensitive to *micromorphic* or substance properties (e.g., roughness, hardness) than to *macromorphic* or shape properties (stereognosis). Likewise, Klatzky, Lederman, and Reed (1987) cited texture and hardness as attributes that are more salient for haptics than vision, and concluded that haptics is oriented towards the encoding of substance and vision towards the encoding of shape (also see Klatzky & Lederman, 1987). Touch dominates vision in judgments of roughness (Heller, 1982, 1989; Lederman, Thorne, & Jones, 1986), and information on properties such as temperature, weight, and hardness is generally available only to haptics (Klatzky & Lederman, 1987).

Katz noted, too, that the importance of touch might be better appreciated if as many people lacked the sense of touch (numbness), as lacked sight (blindness) or hearing (deafness). At the same time, he pointed out commonalities between the senses. "Evolutionally the vibratory sense represents the bridge between pressure sense and auditory sense. In deaf people it functions vicariously for the absent auditory sense, and may even render music accessible to them" (Katz, 1952, p. 203). Also, he generalized the three spatial modes of appearance of color (film, surface, volume) to touch (Sections 6 and 7).

Touch Redux

Echoing Katz's point on the richness of touch, Kennedy (1978) noted that touch provides information on viscosity, slipperiness, softness, texture, and elasticity, and he said that "the possible variations in resistance are so myriad, so complex that they defy any standard physics of location and pressure, any ordinary-language categorizing schemes . . ." (p. 294). Gibson (1966) listed such tactual experiences as stroking, caressing, twisting, pulling, sucking, prodding, and "the crawling insect, the scratching thorn, the brushing leaf, and the shaking branch that need to be distinguished" (p. 117). He reported that blindfolded subjects can discriminate, without error, between stroking devices that rub, scrape, brush, or roll on the skin.

Katz's other classic volume, *Farbwelt* (World of Color), first appeared in German in 1911, was revised and enlarged in 1930, and then appeared in English translation in 1935 (Katz, 1911, 1935). Robert D. MacLeod (personal communication, April 1970), who translated *Farbwelt* into English, said:

> "Katz considered the *Tastwelt* more important than the *Farbwelt*. To some extent he was probably correct. Both are great pioneering works, but the *Tastwelt* explores a field that has never been properly cultivated and that each year gains in significance. The current interest in the body percept, for instance, would be enlightened by a reading of the *Tastwelt*. Katz had a genius for seeing problems in everyday phenomena."

Gibson (personal communication, May 1968) said of *Tastwelt*: "The book has been neglected. I owe more to it than I have recognized recently. I had forgotten that he challenged Johannes Mueller, for example.... That was a bold stroke."

Katz, like Gibson, emphasized the importance of higher-order invariants in the perception of objects, and the holistic quality of perception in time as well as space. In Sections 2 and 14, he deplored the typical atomism, with respect to both time and space, in the sensory psychology of touch, an atomism that bespoke of a "tachistoscopic" mentality. He acknowledged the yeoman service of the physiologists in delineating the peripheral receptors for touch, warm, and cold, but said that psychologists ought to be studying an entirely different realm of complex phenomena. He felt that the early, intimate association between psychologists and physiologists in the tactile sensory area had stunted the independent development of the psychological side. He applauded the holistic concerns of such physiologists as Weber and Sherrington.

The Intelligent Hand

Both Katz and Gibson emphasized movement, and both pointed out how the hand can be wielded in active touch, so that stimulation is obtained, rather than imposed (see also Heller, 1984, on the advantage of active over passive touch). "The hand can grope, palpate, prod, press, rub, or heft..." (Gibson, 1966, p. 123). Katz regarded the hand, not the minute receptors that the physiologists dig out of it, or the entire skin surface, as the organ of touch. He emphasized the role of the hand as a unitary sense organ, noting that Kant called the hand man's outer brain (Sections 1 and 12). Likewise, Klatzky and Lederman (1987) and Lederman and Klatzky (1987) described the exploratory procedures of "the intelligent hand," i.e., the stereotypical hand movements, each associated with the extraction of specific object properties; their work supports Katz's emphasis on the importance of purposive touch in exploration. The general conception of a sense organ is of something quite compact and unitary (e.g., eye, ear, nose, mouth; Gibson, 1966). Katz's bold stroke was to propose an organ, the hand, that is quite as unitary as the organs of the other senses.

Katz was impressed by the high degree to which the information received by the moving hand enables the person to discern the surface qualities (roughness, hardness, graininess, etc.) and thus the specific type of material or substance being felt. Whereas in vision, movement generally impedes perception of surface qualities, in touch it is *lack* of movement that is most damaging (Section 17). The touch organ, when kept motionless relative to the object, is beset with a partial anesthesia. Movement is well-nigh as indispensable to touch as light is to vision.

Vibration Sense

For Katz, movement of the touch organ relative to the object touched was crucial (Sections 14 to 17) in large part because it provided input for the sense of vibration (Sections 39 to 45), which Katz regarded as an important and sensitive modality, and as one which is independent of the sense of pressure (Section 41). Whereas the pressure sense is a proximal sense, the vibration sense is also a remote sense; it helps us to detect the approach of a far-off train or a herd of buffalo (Section 43).

Katz (1937a) reviewed some very interesting research by himself and his associates on the ability to localize vibratory sources beyond the body's surface. Normal and deaf subjects who touched a dinner plate with their hands could accurately report where on the plate a vibrator was applied. In further tests, subjects placed their fingers or hands at various distances along a wooden bar and reported whether the vibrator was applied on the left or right side of the bar. Even with a single hand or with the two hands crossed (i.e., the left hand was placed on the right side), the vibration could be properly localized. A slightly earlier and slightly stronger stimulation on one side of the hand prompted the subject to localize the source towards that side. Given that vibrations travel through wood at five km/sec, the time difference could not have exceeded a fraction of a msec. Using separate vibrators for the two hands confirmed that localization could be based on either a time difference or an intensity difference. More recently, Gescheider (1974) reported on further tests in which a time difference was traded against an intensity difference in determining the apparent localization (see also von Békésy, 1967).

The pressure sense determines the presence of a surface, whereas the vibration sense determines its properties, according to Katz. Merely resting the hand on a material may evoke the impression of a surface, but to discern its surface qualities, and to recognize its specific material, movement is necessary. Katz's subjects wiped, hefted, poked, scraped their nails on, and chewed various materials in order to identify them. His subjects could even discriminate fairly well among such materials as wood, porcelain, metal, and paper by hitting them with a hammer. Even when the hammer made contact with an iron plate for barely 3 to 5 msec, the blow and the resulting vibrations nevertheless produced confident recognition of the material.

Katz found that adaptation occurs quickly for the motionless touch organ, but not when movement is present (Section 15). He concluded that the vibratory sense, like the auditory sense, is almost tireless (Katz, 1930). Other evidence indicates, however, that considerable adaptation occurs even with the vibration sense (Geldard, 1940a, 1940c; Hahn, 1974; Lederman, Loomis, & Williams, 1982). Also, Walker (1967) found evidence for textural aftereffects involving roughness. If coarse sandpaper is stroked for several seconds by fingers of one hand, and fine sandpaper by fingers of the other hand, then an intermediate grade sandpaper will feel finer to the fingers of the first hand and coarser to the fingers of the second hand.

For Katz, the ability of deaf persons to "listen" to music through skin vibrations further attested to the close connection between skin and ear; for more recent research tracing similarities between ear and skin, see von Békésy (1967). Katz's evidence is suspect, however, because some sense of hearing might yet have been involved in these cases. In a person suffering from conduction or transmission deafness rather than nerve deafness, the intense vibrations produced in concert halls might well reach an intact inner ear through bone conduction (von Békésy, 1967). However, in Section 40, Katz cites Ewald's finding that even with bilateral loss of the labyrinth, deaf persons still respond to many auditory stimuli. Katz proposed that echolocation in bats and the human blind was based on both auditory and vibratory sensitivity (Addendum to Section 43), but later work has shown that the human blind perceive obstacles solely on the basis of auditory information (Worchel & Dallenbach, 1947).

Katz became engaged in a long, spirited debate with von Frey as to whether vibration ought to be considered a separate sense from pressure (Section 41). For Katz, "tactual" proper meant the four basic cutaneous qualities or specific nerve energies proposed by von Frey (pressure, pain, warm, cold; Boring, 1942), plus a fifth (vibration) which he added himself. (Defined more broadly, the "sense of touch," or haptics, encompasses both cutaneous and kinesthetic or muscle-joint-tendon sensitivity: Gibson, 1963,

1966; Gordon, 1978; Loomis & Lederman, 1986; Sections 2 and 48 of the World of Touch.) Von Frey, by contrast, credited the pressure sense with our impressions of vibration, as well as those of contact, pressure, and tickle. Katz said that vibration is a separate modality, but that it need not have its own sense organs. He was inclined to agree with von Frey's (1904) view that the vibration sensations are tied to the organs of the pressure sense, but also held, contrary to von Frey, that these organs can transmit vibratory and pressure sensations simultaneously and independently. Vibration, then, is no mere off-shoot or variant of the pressure sensation, but something qualitatively different. Katz thus denied the validity of Mueller's principle of specific nerve energy, according to which each nerve produces a specific conscious quality when excited (Section 44). (Verrillo, 1968, cites other investigators of the skin senses who dissented from Mueller's position.)

Katz presented several examples to demonstrate that felt vibration and pressure differ. For example, touching with a stick held between the teeth (Section 26) or with the fingernail rather than the fingertip (Section 24), which presumably excludes the pressure sensation, still leads to good recognition of the material. Katz also made much of the fact that the tongue is very sensitive to pressure, but very insensitive to vibration (Sections 41 and 45; see also Katz, 1936); he attributed the latter to the weak resonance of the tongue. However, Geldard (1940c) was puzzled as to "how the myth of lingual vibratory insensitivity ever got started" (p. 294), since several studies have shown that the tongue is highly sensitive to vibration. If the tongue's vibratory sensitivity lags somewhat behind its tactual sensitivity, "the answer lies in its lowered capacity to conduct forced vibrations" (p. 294).

Geldard (1940a, 1940c) reviewed the controversy between Katz and von Frey, and related matters. Based on his own data and those of others, which indicated that the skin spots most sensitive to pressure are the ones most sensitive to vibration as well, Geldard (1940b, 1940c) concluded that there is not a separate vibratory sense. However, Katz readily conceded that pressure and vibration may share the same peripheral receptors, so this evidence does not seem to refute his position.

Although Katz tended to agree with von Frey that vibration and pressure share a common set of peripheral receptors on the skin, he withheld making a final judgment. Recently, the Pacinian corpuscle has been identified as the sensory structure responsible for the detection of vibration (Quilliam, 1978; Vallbo & Johansson, 1978). The Pacinian is a large, ovoid corpuscle, having onion-skin coverings, which projects to the spinal cord, and possibly the brain, via a single, direct fiber. Ironically, the receptor for the less controversial pressure sense has not yet been identified. "With one notable exception (i.e., the Pacinian corpuscle), it is not yet known for certain to what kinds of physical stimuli cutaneous receptors will respond" (Quilliam, 1978, p. 11). In addition to the Pacinian system, which is most sensitive to high-frequency vibrotactile stimulation (100-300 Hz), there also is evidence for a non-Pacinian system that is sensitive to low-frequency vibrotactile stimulation, particularly below 40 Hz (Talbot, Darian-Smith, Kornhuber, & Mountcastle, 1968; Verrillo, 1968). According to Talbot et al., receptors in the cutaneous tissue respond to low-frequency stimulation, whereas the subcutaneous Pacinian system responds to high-frequency stimulation; low-frequency stimulation is felt as a light flutter on the skin that can be localized accurately, whereas high-frequency stimulation is felt as a deeper-lying vibratory hum or buzz that is less accurately localized. Some data indicate that two distinct populations of non-Pacinian receptors are sensitive to vibrotactile stimulation in addition to the Pacinian system (Capraro, Verrillo, & Zwislocki, 1979; Gescheider, Sklar, Van Doren, & Verrillo, 1985).

Texture

Katz dealt with a wealth of tactual information, but above all, he was concerned with texture or *microstructure*, that is, the fine structure of the surface, rather than its shape or *macrostructure* (Sections 4, 9, and 10). The microstructure reveals the qualities (*Modifikationen*), which lead to the classification of a material on such dimensions as rough-smooth (Sections 18 to 28). It also reveals the identifying characteristics (*Spezifikationen*), which inform us as to whether the material is paper, leather, metal, etc. (Sections 29 and 30). Katz considered the texture or microstructure to be independent of the shape or macrostructure; no matter how a piece of wood is carved, for example, it keeps the same grain or texture.

In his classic volume on vision, Katz (1911, 1935) showed perhaps an even greater concern for microstructure. A similar emphasis is evident, too, in Gibson's (1950) analysis of visual texture gradients. Gibson (1966) said that the visual "texture of a surface is probably even more important to animals than its pigment color in identifying it . . ." (p. 126). In Section 9, Katz said that "color can deceive, but texture cannot do so as easily." More recent studies confirm that textural tasks are performed very well by the haptic system (e.g., Heller, 1982, 1989; Morley, Goodwin, & Darian-Smith, 1983; Lamb, 1983), and that texture is a highly salient property for touch (Klatzky, Lederman, & Reed, 1987; Lederman, Thorne, & Jones, 1986).

Katz's concern with microstructure (ground) goes against the grain of Gestalt psychology, in which the form and its contours (figure) seem largely to pre-empt other concerns. Nuances, textures, and meanings may attach to events that involve holistic perception, but do not properly fall under the rubric of form. "All forms are wholes, but . . . not all wholes are forms" (Katz, 1950, p. 39). Katz's holism was more thorough-going than that of the

Gestalt psychologists, too, in that he gave equal consideration to the temporal and spatial domains. He attacked atomism in both time and space. His concern with active touch, movement, and vibration attested to his own commitment to temporal holism. Katz (1950) noted that the Gestalt psychologists, while they professed to oppose time atomism (e.g., in movement, melody), turned readily to the tachistoscope to obtain evidence concerning form-creating processes (see also Section 14).

Texture is revealed most fully by the vibration sense, but not in a simple fashion. Moving the hand faster produces a higher "pitched" vibration, but people apparently can take both "pitch" and hand speed into account, and thus obtain an invariant impression of a material. Katz found (Section 45) that judgments on roughness remained essentially the same over a 10-fold range in speed (1 to 10 cm/sec). Thus, Katz discovered an instance of roughness constancy. Lederman (1974) likewise found little change in felt roughness over a 25-fold range in speed (1 to 25 cm/sec). Like Katz, she found that a material feels slightly smoother at a higher rate of hand movement.

Katz found that felt roughness increased, however, when materials were moved across the stationary fingers with increased force (Section 23). Lederman (1974, 1978) similarly found that felt roughness increased when the fingertip force exerted on a grooved plate or piece of sandpaper increased. Katz demonstrated that the increased felt roughness was not due to increased resistance of the hand to the movement rather than stronger vibrations. He ran his hand down a composite sheet (two sheets of different texture glued back-to-back), with the thumb on one side, and the index and middle fingers on the other. He clearly perceived two different impressions at the same time. Taylor and Lederman (1975), too, showed that the resistance felt in the hand is irrelevant to the judgment of roughness. They lubricated a grooved aluminum plate with liquid detergent to reduce the coefficient of friction and thus the resistance felt in the hand, but found no change in felt roughness when subjects moved their fingers across the soaped plate rather than a dry one.

Katz obtained evidence that roughness discrimination was not dependent solely on pressure sensations (Section 24). Papers could be discriminated rather well on roughness even when touched by the fingernail rather than fingertip, or with a stick held in the hand or teeth. Also, a smooth glass surface could be discriminated from an etched glass surface in which the differences in level (bumps) were far less than .001 mm; pressure sensations are produced by larger bumps, and feed into the spatial sense of the skin. However, although Katz thus ruled out the pressure sense as the main basis for felt roughness, he did not thereby rule in the vibration sense. True, he reported that the fingers could be "deafened" to roughness by cooling them (see also Green, Lederman, & Stevens, 1979) or by coating them with a plastic (collodion) film or a sticky glue, which precluded vibration sensations. As Lederman (1982) pointed out, however, such procedures may not only eliminate vibration sensations, but also affect other key factors, such as lateral "shearing" forces and skin flexibility. Katz reported that continuous noises differing in intensity and quality could be heard near surfaces receiving touching movements, and that the entire hand and even the forearm resonated to certain touching movements (Section 45), but he provided no evidence that this vibratory input is sufficiently rich to account for the roughness judgments (see also Lederman, 1979). A third, as yet unspecified system may mediate roughness judgments; the choice need not be between a purely spatial system (pressure) and a purely temporal system (vibrations).

Vibration presumably depends on spatial frequency or density, and thus ought to be a joint function of element size and element spacing, but recent studies by Lederman and her associates using linear gratings have found that felt roughness depends mainly on element spacing (see Lederman's, 1982, review). Furthermore, Lederman, Loomis, and Williams (1982) found that prior vibratory stimulation attenuated felt vibration, but not felt roughness. They concluded, contrary to Katz, that roughness is encoded based on interelement spacing and force applied to the skin, not on temporal pulse frequency. However, felt roughness may not have depended on the vibratory sense in Lederman's studies because the linear gratings she used generally had fairly large bumps; e.g., the smallest spatial period (1 groove + 1 ridge) that Lederman, Loomis, and Williams used was .63 mm. Katz conceded (Section 45) that if differences in surface level exceeded a certain threshold (which he thought clearly occurs with coarse fabrics), then they are recognizable to the sensory organs for pressure (see also Heller, 1989). "I do not attribute all roughness judgments to vibration sensations, but call upon the latter only for the levels of roughness that are no longer apparent to the spatial sense of the skin. . . ."

In Section 11, Katz said that there is nothing comparable to a magnifying glass or microscope to improve one's touch. Typically, an intermediary, such as gloves, somewhat "deafens" the fingers (Section 24), although making touching movements tends to increase the touch transparency of the intervening medium (Section 8). In Section 30, Katz noted that touching with gloves would preclude or attenuate the recognition of the substance, but not the form of an object. Thus, the use of gloves provides a way to determine whether pathological astereognosis (tactual object blindness) is due more to a failure in the recognition of substance or a failure in the recognition of form. However, an intervening cloth or glove may sometimes aid touching, perhaps by amplifying or magnifying the sensory input (Lederman, 1978). Katz reported that one blind person could read Braille text very well through kid leather gloves, and that some blind girls tried to make the task easier by reading the Braille text through an apron (Section 32). The procedure that craftsmen such as carpenters (Kennedy, 1978) and automobile body inspectors (Lederman, 1978, 1982) use to examine the finish on surfaces—wiping the surface with a paper or rag, or while wearing a cotton glove—has recently been studied (Gordon & Cooper, 1975; Lederman, 1978). Gordon and Cooper found that the orientation of a surface undulation is better detected when one runs a thin, intermediate paper across a surface, rather than the bare fingers. Gordon and Cooper said that the thin intermediate paper or cloth may improve performance by turning off the light-pressure system which produces felt roughness and which might normally mask the deeper receptors. The intervening layer also ought to reduce or eliminate the effect of perspiration; that, too, ought to decrease the felt roughness, as Katz found when he applied a trace of drying powder to the fingers (Addendum to Section 36). However, Lederman (1978) found that the intermediate paper increased rather than decreased the felt roughness. Lederman said that the paper may be an amplifier, not an attenuator, because its edges catch on surface irregularities. Perhaps the paper helps because the skin itself is too smooth to "catch" on small surface irregularities. If so, then a rough paper ought to work better than a smooth paper. Lederman (personal communication, November, 1979) said that the rougher (non-glossy) side of the intermediate paper does indeed work better, and that she always uses that side in her tests.

The tactual judgment also might be aided by the sounds produced when an object or substance is touched (Sections 18, 26, 44). To eliminate such auditory cues, Katz typically stopped the ears of his subjects, although he doubted whether this was totally effective in eliminating auditory cues. Oddly, Katz never systematically investigated the effect of auditory cues. More recently, Lederman (1979) found that magnitude estimates of roughness did not change when subjects' ears were stopped, which indicates that her subjects tended to use only tactual cues when both auditory and tactual cues were present. Somewhat different (and more variable) magnitude estimates of roughness were obtained, however, when tactual cues, rather than auditory cues were eliminated (by having subjects listen to the experimenter touch the grooved plates). The greater variability of the auditory-based estimates may explain why subjects tended to rely more on tactual than auditory cues. Subjects' reports indicated that they used both pitch and loudness as auditory cues to roughness.

Katz considered the temperature sense, not the pressure sense, to be the second most important source of tactual information, after the vibratory sense (Sections 36 to 38). This reveals again that Katz was more concerned with an object's substance or material than its shape or contour. A material's ability to absorb heat represents the same kind of invariant property as its ability to reflect light (reflectance or albedo). Metals feel positively cold.

And wools feel positively warm, even though wool must rise to the body temperature if it is initially cooler. Temperature sensitivity aids more in identifying characteristically cold (vs. warm) materials, perhaps because such materials are relatively rare. Katz might have considered other explanations as well. Cold materials might be more salient because they deviate more from the neutral point than do warm materials, owing to the great ability of the metals to conduct heat away from the skin surface, or because there are nearly ten times as many cold spots as warm spots on the skin (von Frey, 1904).

Katz's Phenomenology

Unlike Gibson, Katz did not want to rid perception of its phenomenological aspect; in the World of Touch, for instance, *phenomenon* refers to the phenomenally experienced percept. True, Katz sometimes emphasized the informational aspect, as when he focused on the aid that vibrations provide in identifying various materials, rather than on the phenomenal impressions they produce. Overall, though, Katz was "one of this century's outstanding exponents of psychological phenomenology" (MacLeod, 1954, p. 1; also see Arnheim, 1953); "phenomenology for him was essentially an attitude of 'disciplined naivete'" (MacLeod, 1954, p. 3). He focused on the phenomenological aspect, for example, when he considered whether a perceived object was projected "out there." Katz, like the Gestaltists, was more mindful than Gibson of the perceiver's contribution to the final percept. In reading Katz's World of Touch, in fact, one is struck first by the great wealth of phenomenological details.

Important influences on Katz were the philosopher Edmund Husserl, whose lectures he attended at Goettingen, and through Husserl the phenomenological movement that was gaining strength in German philosophy. In his autobiography, Katz (1952) said: "None of my academic teachers with the exception of G. E. Mueller has more deeply influenced my method of work and my attitude in psychological matters than Husserl by his phenomenological method" (p. 194). Husserl was concerned with the pure conscious act, with its immediate felt experiencing and directness or intentionality; "Husserl derived the notion of 'intentionality' from his mentor Brentano" (Jennings, 1986, p. 1236). Husserl's student, Martin Heidegger, united the existentialism of Kierkegaard and Nietzsche with the phenomenology of Husserl, in his most important work, *Being and Time.*

According to MacLeod, Katz insisted that the psychologist should deliberately suspend "his physical, physiological, and philosophical biases and attempt to observe phenomena as they are actually presented. The phenomenal world thus viewed contains properties and relationships that escape the notice of the physically or physiologically oriented observer. The classical psychologist was content to order colors in terms of hue, brightness, and saturation; Katz saw them also varying in mode of appearance, pronouncedness, insistence, transparency, inherence, and stability. Classical psychology was busily mapping the patterns of pressure, pain, warm, and cold spots on the skin, and searching for receptors; Katz went further, and explored the active process of 'touching' (*tasten*), discovering here, too, modes of appearance, properties of organization, and unsuspected kinds of sensitivity. It is unfortunate that, while his visual studies have been widely appreciated, his richly suggestive book on the world of touch has received relatively little notice" (MacLeod, 1954, p. 3). The phenomenological distinctions Katz made between the subjective and objective poles (Section 4) and between the modes of appearance of touch (immersed touch, surface touch, and volume touch; Sections 6 and 7) will now be discussed in turn.

Subjective and objective poles. For Katz, the projection of the object outside the body distinguished the objective pole or side (e.g., "I feel a pointed object out there"), in which attention is directed to the distal object projected "out there," from the subjective pole or side (e.g., "I feel a prickling sensation"), in which attention is directed to the proximal stimulus. Typically, the objective mode, with its external projection, dominates. Thus, the impression of hardness or softness usually is felt at the writing point of the pencil rather than in the hand holding the pencil. Likewise, an automobile driver feels the goodness of the road, and an airplane pilot feels the elastic qualities of the air rather than merely the local jiggling of a steering wheel or rudder in the hand (Section 25). Katz detailed conditions in which the subjective mode is favored instead, such as when an object is impressed upon a passive touch organ (Section 4). Modality also is very important. The objective side especially predominates in the case of vision, whereas pain seems universally to be regarded as subjective; even Gibson (1963) acknowledged that one may say that a stomach-ache is sensory rather than perceptual.

The objective side has provided the prototype or ideal for perception in general ever since Thomas Reid (1764, 1785) distinguished between sensation and perception. According to Reid, sensation precedes perception and provides the information upon which perception is based, and the object is projected beyond the receptive surface in perception, but not in sensation. It is now generally accepted, however, that perceivers do not have direct access to sensations, if such should exist, but rather must tease out the subjective pole or side via perception (Boring, 1952; Gibson, 1963). For Gibson (1950, 1952), like Katz and others, the subjective side (visual field) was an *alternative* experience, not simply the basis for the other experience. Gibson's objective side (visual world) corresponds with the "rigid, Euclidean, natural, tape-measured world" (Boring, 1952, p. 146), whereas the subjective side (visual field) tends to be "bidimensional, pictorial, and in a sense 'anatomical' like the retinal image" (Boring, 1952, p. 144). The visual field, however, is never fully achieved: It "is never flat, . . . it is never wholly depthless. Nor is it lacking in the character of being *outside* of us, in externality. Nevertheless, it has *less* of these qualities than the visual world" (Gibson, 1950, p. 42).

Many other psychologists likewise have been sensitive to the bimodal nature of perception (e.g., Brunswik, 1956; Mack, 1978; Rock, 1977). The perceiver may attend either to the distal stimulus (distal or constancy mode) or to the proximal stimulus (proximal mode) (Mack, 1978; Rock, 1977). The two modes may sometimes alternate in rapid succession, as, for example, when railroad tracks are seen almost simultaneously as parallel and as converging (Mack, 1978). In vision, the objective aspect normally predominates, and the perceiver must take a special attitude, like that of the artist, or reduce the stimulus input (e.g., by presenting a homogeneous, blurred visual field; or by excluding most of the visual field with a reduction screen) in order to obtain the basic subjective impressions. In other cases, though, little or no effort may be needed (e.g., pains, after-images).

In the objective pole, the percept not only is projected beyond the body surface, but also beyond the time of sensory input, and there is a diminution in the richness of modality-specific qualities (Krueger, 1982). Thus, in vision, where the objective pole predominates, an object does not cease to exist when the illumination is removed (Katz, 1935). An object brought into contact with the skin produces two different impressions at the same time: 1) the impression of an object that had pre-existed the contact, and 2) the impression of a contact that was suddenly created (Michotte, 1950). This observation fits Katz's view that the objective and subjective poles are both available in touch.

To transcend the particular accidents of space, time, and modality quality in the objective pole means to render these dimensions transparent to the "pure information" available in the external input. Gibson, especially, emphasized the objective pole. "For Gibson, the information we acquire perceptually is . . . amodal" (Natsoulas, 1978, p. 280). Thus, the fact that wielding an object in different ways (tossing, shaking, etc.) usually produces a unitary impression, indicates that "the merely proprio-specific information . . . [is] filtered out, as it were, leaving pure information about the object" (Gibson, 1966, p. 127). The permanent properties of the object, the invariants over time, are extracted from the sensory input, and the perceiver "ordinarily pays no attention whatever to the flux of changing sensations" (Gibson, 1966, p. 3).

Gibson was mainly concerned with vision, where the objective pole predominates. The subjective side is more evident in other senses, such as touch. Thus, Katz noted that an impression of movement may always accompany that of elasticity; something elastic cannot be imagined without also imagining a movement being carried out on an elastic object (Section 16). By and large, though, Katz shared Gibson's emphasis on "pure information" and the objective pole. Movement need not be a conscious feature of the impression of stickiness, he said. Even so simple a quality as roughness bears no trace of movement in its contents (Section 16). In the impression of the ordinary object, according to Katz, movement, time, and space leave no trace of themselves; the object is precipitated as an independent entity, largely uncontaminated by its journey through tactual time and space. When a person glides his or her hand over a motionless object, say the corner of a chair, the hand touches different and constantly changing parts of the corner, Katz said, yet the corner persists just as steadfast in tactual space as do objects in visual space when the eye is moved (Section 17).

Movement favors the objective pole, according to Katz. For example, when the hand is moved to touch another part of the body, an objective impression obtains in the moving hand, but a subjective impression obtains in the motionless area being touched. This might explain why it is difficult to tickle oneself (Weiskrantz, Elliott, & Darlington, 1971); an objective tickle "out there" becomes a mere subjective touch on the foot when your own hand and arm control the tickling strokes. The reafferent signal arising from the self-produced movement may be cancelled out in this case (Weiskrantz et al.), just as it is when the hand is moved over the parts of a motionless chair.

For Gibson (1962), too, active movement favors the objective pole, which he termed object-form (vs. skin-form). Thus, pressing a finger into something reveals the object's qualities, whereas impressing the object upon a passive touch organ produces labile sensations referred to the skin. An active grasp might be needed, for instance, to convert a pressure sensation into the impression of hardness. Similarly, William James (1890) noted that two contacts are felt if something is placed between two fingers, but a single object is experienced if the fingers are squeezed together.

Modes of appearance: Immersed touch, surface touch, and volume touch. Katz indicated how objects might be cohered on the objective side when he described touch analogs to the three primary spatial modes of appearance of color (film, surface, volume). The objective pole predominates to varying degrees in all of these modes, since objects, materials, and space are projected "out there." Conceiving of a touch analog to surface color is not difficult; when one touches an object made of some solid substance (e.g., metal, glass, wood) and having a continuous, unbroken surface, one experiences a definite surface at a definite distance and definite orientation. But what touch impression would correspond to film color, whose misty, spongy appearance, produced by homogeneous illumination, contrasts sharply with the hard, impenetrable, and definitely localizable appearance of surface color? A strong, rapid stream of air or liquid, Katz said, produces a space-filling immersed touch (*raumfuellendes Tastquale*). The stream feels indeterminate as to form; there may be a suggestion of a certain thickness, but the "form" lacks a rear boundary and is always perceived as lying in the frontoparallel plane. Immersed touch may arise in drawing the hand through water or a thick liquid (Katz, 1936), or in brushing cobwebs aside (Kennedy, 1978). Katz said that "the moving atmosphere strikes us subjectively as an immersed touch. This may have prompted [so-called] primitive people to make animistic interpretations of inanimate natural processes" (Section 6).

Immersed touch does not represent the qualities of a body; it characterizes a substance, not an object. In immersed touch, the resistance that the material offers the hand is experienced as elastic rather than stiff or rigid. Immersed touch and film color belong to the objective pole, but are more subjective than surface touch and surface color. To achieve a subjective attitude, Katz (1935) recommended that painters try to see colors as filmy. Katz said that moving the fingers through loose sand or flour produces neither a surface touch nor an immersed touch (Section 10), but I would submit that the impression comes closer to that of an immersed touch. Also related to immersed touch and film color is the touch-transparent film of a loose glove or a piece of paper felt between the fingers (Sections 8, 31), and the wetness, oiliness, or stickiness through which a surface is felt (Section 9).

Katz (1937b) studied what bakers meant when they used terms such as good body, good spring, lively, good elasticity, and clay-like to describe flour. He found that good body means that a dough has a minimum of stickiness. The "general feeling of unpleasantness towards sticky things is very likely to be an important factor in the tendency of the baker to make his dough as little sticky as possible" (p. 389). He did not report, though, whether using sticky dough produces bread that is any less appetizing after it is baked! Katz thought that the difference between lively and clay-like or dead was based upon the coordination of antagonistic muscles. After the fingers move together and touch, there is a short opposite movement or bounce. An elastic dough will help this reflex movement, but not a dough of poor elasticity. One feels full of life, the other feels dead.

Katz and Stephenson (1938) found that a weight on a spring feels just as heavy as a free weight weighing only 60% as much. Thus, an elastic pull of 4 kg feels only as heavy as a 2.5-kg dead weight. Perhaps an object on an elastic band feels lighter because it appears to be larger or more voluminous, and less definitely localizable, than an object on a fixed-length string. Thus, Katz may have uncovered a variant of the size-weight illusion, in which a more voluminous material, e.g., a pound of feathers, feels lighter than a denser or more concentrated material, e.g., a pound of iron.

When a solid object is felt through a soft material, the latter seems to fill the intervening space between the hand and the object, producing a space-like volume touch (*raumhafte Tastphaenomene*). The solid object must have sharp contours; putting the blanket on a flat surface, such as the floor, does not produce volume touch. The perceiver may direct his or her attention to the object itself, as occurs when the physician feels an organ through the skin by means of palpation and percussion, and ignores the voluminous feel of the intervening tissue.

Katz (1930, 1936, 1937a) devised a "percussion phantom" to help train medical students in the art of percussion. The students tapped a cardboard square placed over an opening in a box, in order to determine the shape of a vibration-absorbing lead plate that had been attached underneath the cardboard. "In general, the thicker the plate, the easier the task" (Katz, 1930, p. 84). Percussion also has been used to test Swiss cheese. "In a first-class cheese the holes must be of the right size and number, and this can be determined by the percussion method" (Katz, 1930, p. 86).

This introduction provides an overview of Katz's World of Touch, and indicates how the work was later extended and elaborated upon, both by Katz himself and others. It can only hint at the richness of the volume, however, and the reader is invited to feast upon the work itself, which is presented next, beginning with the author's preface.

Lester E. Krueger

References

Arnheim, R. (1953). David Katz, 1884-1953. *American Journal of Psychology, 66,* 638-642.
Békésy, G. v. (1967). *Sensory inhibition.* Princeton: Princeton University Press.
Boring, E. G. (1942). *Sensation and Perception in the History of Experimental Psychology.* New York: Appleton-Century-Crofts.
Boring, E. G. (1952). Visual perception as invariance. *Psychological Review, 59,* 141-148.
Brunswik, E. (1956). *Perception and Representative Design of Psychological Experiments.* Berkeley: University of California Press.
Capraro, A. J., Verrillo, R. T., & Zwislocki, J. J. (1979). Psychophysical evidence for a triplex system of cutaneous mechanoreception. *Sensory Processes, 3,* 334-352.
Geldard, F. A. (1940a). The perception of mechanical vibration: I. History of a controversy. *Journal of General Psychology, 22,* 243-269.
Geldard, F. A. (1940b). The perception of mechanical vibration: II. The response of pressure receptors. *Journal of General Psychology, 22,* 271-280.

Geldard, F. A. (1940c). The perception of mechanical vibration: IV. Is there a separate "vibratory sense"? *Journal of General Psychology, 22,* 291-308.
Geldard, F. A. (1953). *The Human Senses.* New York: Wiley.
Geldard, F. A. (Ed.) (1974). *Conference on Cutaneous Communication Systems and Devices.* Austin, Texas: Psychonomic Society.
Gescheider, G. A. (1974). Temporal relations in cutaneous stimulation. In F. A. Geldard (Ed.), *Conference on Cutaneous Communication Systems and Devices* (pp. 33-37). Austin, Texas: Psychonomic Society.
Gescheider, G. A., Sklar, B. F., Van Doren, C. L., & Verrillo, R. T. (1985). Vibrotactile forward masking: Psychophysical evidence for a triplex theory of cutaneous mechanoreception. *Journal of the Acoustical Society of America, 78,* 534-543.
Gibson, J. J. (1950). *The Perception of the Visual World.* Boston: Houghton Mifflin.
Gibson, J. J. (1952). The visual field and the visual world: A reply to Professor Boring. *Psychological Review, 59,* 149-151.
Gibson, J. J. (1962). Observations on active touch. *Psychological Review, 69,* 477-491.
Gibson, J. J. (1963). The useful dimensions of sensitivity. *American Psychologist, 18,* 1-15.
Gibson, J. J. (1966). *The Senses Considered as Perceptual Systems.* Boston: Houghton Mifflin.
Gibson, J. J. (1979). *The Ecological Approach to Visual Perception.* Boston: Houghton Mifflin.
Gordon, G. (Ed.) (1978). *Active Touch. The Mechanism of Recognition of Objects by Manipulation: A Multi-disciplinary Approach.* Oxford: Pergamon Press.
Gordon, I. E., & Cooper, C. (1975). Improving one's touch. *Nature, 256,* 203-204.
Green, B. G., Lederman, S. J., & Stevens, J. C. (1979). The effect of skin temperature on the perception of roughness. *Sensory Processes, 3,* 327-333.
Hahn, J. F. (1974). Vibratory adaptation. In F. A. Geldard (Ed.), *Conference on Cutaneous Communication Systems and Devices* (pp. 6-8). Austin, Texas: Psychonomic Society.
Heller, M. A. (1982). Visual and tactual texture perception: Intersensory cooperation. *Perception & Psychophysics, 31,* 339-344.
Heller, M. A. (1983). Haptic dominance in form perception with blurred vision. *Perception, 12,* 607-613.
Heller, M. A. (1984). Active and passive touch: The influence of exploration time on form recognition. *Journal of General Psychology, 110,* 243-249.
Heller, M. A. (1989). Texture perception in sighted and blind observers. *Perception & Psychophysics, 45,* 49-54
James, W. (1890). *The Principles of Psychology,* Vol. 2. New York: Holt.
Jennings, J. L. (1986). Husserl revisited: The forgotten distinction between psychology and phenomenology. *American Psychologist, 41,* 1231-1240.
Katz, D. (1911). Die Erscheinungsweisen der Farben und ihre Beeinflussung durch die individuelle Erfahrung (Modes of appearance of color and their modification through individual experience). *Zeitschrift fuer Psychologie,* Ergaenzungsband 7 (Supplement Volume 7), 425 pp. Leipzig: Johann Ambrosius Barth. (A revised and enlarged edition was published in 1930 as *Der Aufbau der Farbwelt.*)
Katz, D. (1920). *Die Erscheinungsweisen der Tasteindruecke* (Modes of appearance of the tactual impression). Rostock: Kommissionsverlag von H. Warkentien.
Katz, D. (1921). *Zur Psychologie des Amputierten und seiner Prothese* (The psychology of

amputees and their prostheses). *Zeitschrift fuer angewandte Psychologie*, Beiheft 25 (Supplement 25), 118 pp. Leipzig: Johann Ambrosius Barth.

Katz, D. (1923a). *Der Vibrationssinn* (The vibratory sense). Scripta universitatis atque bibliothecae Hierosolymitanarum. The Hague: Kommissionsverlag M. Nijhoff.

Katz, D. (1923b). Ueber die Natur des Vibrationssinnes (On the nature of the vibratory sense). *Muenchener medizinische Wochenschrift*, No. 22.

Katz, D. (1925). Der Aufbau der Tastwelt. *Zeitschrift fuer Psychologie*, Ergaenzungsband 11 (Supplement Volume 11), 270 pp. Leipzig: Johann Ambrosius Barth. Reissued in German in 1969 by Wissenschaftliche Buchgesellschaft in Darmstadt.

Katz, D. (1930, May). The vibratory sense and other lectures. University of Maine Studies, Second Series, No. 14. *The Maine Bulletin, 32* (No. 10), 1-163.

Katz, D. (1935). *The World of Colour*. (Translated by R. B. MacLeod & C. W. Fox.) London: Kegan, Paul, Trench, Trubner & Co. (Reissued by Johnson Reprint, a subsidiary of Academic Press.)

Katz, D. (1936). A sense of touch: The technique of percussion, palpation and massage. *British Journal of Physical Medicine, 11* (Old Series), 146-148.

Katz, D. (1937a). Methoden zur Untersuchung des Vibrationssinns (Methods for investigating the vibration sense). In E. Abderhalden (Ed.), *Handbuch der biologischen Arbeitsmethoden* (Handbook of biological methodology) (pp. 879-918). Abt. 5, Teil 7, II (Division 5, Part 7, Chapter II). Berlin: Urban & Schwarzenberg.

Katz, D. (1937b). Studies on test baking. III. The human factor in test baking. A psychological study. *Cereal Chemistry, 14*, 382-396.

Katz, D. (1950). *Gestalt Psychology: Its Nature and Significance*. (Translated by R. Tyson.) New York: Ronald Press.

Katz, D. (1952). Autobiography. In E. G. Boring, H. S. Langeld, H. Werner, & R. M. Yerkes (Eds.), *A History of Psychology in Autobiography*, Vol. IV (pp. 189-211). Worcester, Mass.: Clark University Press.

Katz, D., & MacLeod, R. B. (1949). The mandible principle in muscular action. *Acta Psychologica, 6*, 33-39.

Katz, D., & Stephenson, W. (1938). Experiments on elasticity. *British Journal of Psychology, 28*, 190-194.

Kennedy, J. M. (1978). Haptics. In E. C. Carterette & M. P. Friedman (Eds.), *Handbook of Perception*, Vol. 8 (pp. 289-318). New York: Academic Press.

Klatzky, R. L., & Lederman, S. J. (1987). The intelligent hand. In G. H. Bower (Ed.), *The Psychology of Learning and Motivation*, Vol. 21 (pp. 121-151). San Diego: Academic Press.

Klatzky, R. L., Lederman, S., & Reed, C. (1987). There's more to touch than meets the eye: The salience of object attributes for haptics with and without vision. *Journal of Experimental Psychology: General, 116*, 356-369.

Krueger, L. E. (1970). David Katz's Der Aufbau der Tastwelt (The world of touch): A synopsis. *Perception & Psychophysics, 7*, 337-341.

Krueger, L. E. (1982). Tactual perception in historical perspective: David Katz's world of touch. In W. Schiff & E. Foulke (Eds.), *Tactual Perception: A Sourcebook* (pp. 1-54). New York: Cambridge University Press.

Lamb, G. (1983). Tactile discrimination of textured surfaces: Psychophysical performance measurements in humans. *Journal of Physiology, 338*, 551-565.

Lederman, S. J. (1974). Tactile roughness of grooved surfaces: The touching process and effects of macro- and microsurface structure. *Perception & Psychophysics, 16,* 385-395.

Lederman, S. J. (1978). "Improving one's touch"... and more. *Perception & Psychophysics, 24,* 154-160.

Lederman, S. J. (1979). Auditory texture perception. *Perception, 8,* 93-103.

Lederman, S. J. (1982). The perception of texture by touch. In W. Schiff & E. Foulke (Eds.), *Tactual Perception: A Sourcebook* (pp. 130-167). New York: Cambridge University Press.

Lederman, S. J., & Klatzky, R. L. (1987). Hand movements: A window into haptic object recognition. *Cognitive Psychology, 19,* 342-368.

Lederman, S. J., Klatzky, R. L., Collins, A., & Wardell, J. (1987). Exploring environments by hand or foot: Time-based heuristics for encoding distance in movement space. *Journal of Experimental Psychology: Learning, Memory, and Cognition, 13,* 606-614.

Lederman, S. J., Thorne, G., & Jones, B. (1986). Perception of texture by vision and touch: Multidimensionality and intersensory integration. *Journal of Experimental Psychology: Human Perception & Performance, 12,* 169-180.

Lederman, S. J., Loomis, J. M., & Williams, D. A. (1982). The role of vibration in the tactual perception of roughness. *Perception & Psychophysics, 32,* 109-116.

Loomis, J. M., & Lederman, S. J. (1987). Tactual perception. In K. Boff, L. Kaufman, & J. Thomas (Eds.), *Handbook of Human Perception and Performance,* Vol. II, Chap. 31 (pp. 1-41). New York: Wiley.

Mack, A. (1978). Three modes of visual perception. In H. L. Pick, Jr., & E. Saltzman (Eds.), *Modes of Perceiving and Processing Information* (pp. 171-186). Hillsdale, N.J.: Erlbaum.

MacLeod, R. B. (1954). David Katz, 1884-1953. *Psychological Review, 61,* 1-4.

Michotte, A. (1950). A propos de la permanence phénoménale faits et théories (Facts and theory on phenomenal permanence). *Acta Psychologica, 7,* 298-322. (Reprinted in A. Michotte, et. al. [Ed.], *Causalité, Permanence et Réaliteé Phénoménales* (Phenomenal Causality, Permanence, and Reality) [pp. 347-371]. Louvain: Publications Univ. de Louvain.)

Morely, J., Goodwin, A., & Darian-Smith, I. (1983). Tactile discrimination of gratings. *Experimental Brain Research, 49,* 291-299.

Natsoulas, T. (1978). Residual subjectivity. *American Psychologist, 33,* 269-283.

Quilliam, T. A. (1978). The structure of finger print skin. In G. Gordon (Ed.), *Active Touch. The Mechanism of Recognition of Objects by Manipulation: A Multi-disciplinary Approach* (pp. 1-18). Oxford: Pergamon Press.

Reid, T. (1970; originally published in 1764). *An Inquiry into the Human Mind.* T. Duggan (Ed.). Chicago: University of Chicago Press.

Reid, T. (1878; originally published in 1785). *Essays on the Intellectual Powers of Man.* Abridged edition with notes from Sir William Hamilton and others. J. Walker (Ed.). Philadelphia: J. H. Butler.

Révész, G. (1950). *Psychology and Art of the Blind.* (Translated by H. A. Wolff.) London: Longmans, Green.

Rock, I. (1977). In defense of unconscious inference. In W. Epstein (Ed.), *Stability and Constancy in Visual Perception: Mechanisms and Processes* (pp. 321-373). New York: Wiley.

Rock, I., & Harris, C. S. (1967, May). Vision and touch. *Scientific American, 216,* 96-104.

Schiff, W., & Foulke, E. (Eds.) (1982). *Tactual Perception: A Sourcebook.* New York: Cambridge University Press.

Talbot, W. H., Darian-Smith, I., Kornhuber, H. H., & Mountcastle, V. B. (1968). The sense of flutter-vibration: Comparison of the human capacity with response patterns of mechanoreceptive afferents from the monkey hand. *Journal of Neurophysiology, 31,* 301-334.

Taylor, M. M., & Lederman, S. J. (1975). Tactile roughness of grooved surfaces: A model and the effect of friction. *Perception & Psychophysics, 17,* 23-36.

Vallbo, A. B., & Johansson, R. S. (1978). The tactile sensory innervation of the glabrous skin of the human hand. In G. Gordon (Ed.), *Active Touch. The Mechanism of Recognition of Objects by Manipulation: A Multi-disciplinary Approach* (pp. 29-54). Oxford: Pergamon Press.

Verrillo, R. T. (1968). A duplex mechanism of mechanoreception. In D. R. Kenshalo (Ed.), *The Skin Senses* (pp. 139-156). Springfield, IL: Charles C. Thomas.

von Frey, M. (1904). *Vorlesungen ueber Physiologie* (Lectures on Physiology). Berlin: Springer. Excerpt reprinted in R. J. Herrnstein & E. G. Boring (Eds.), *A Source Book in the History of Psychology* (pp. 49-58). Cambridge, Mass.: Harvard University Press.

Walker, J. T. (1967). *Textural aftereffects: Tactual and visual.* Unpublished doctoral dissertation, University of Colorado, Boulder.

Weiskrantz, L., Elliott, J., & Darlington, C. (1971). Preliminary observations on tickling oneself. *Nature, 230,* 598-599.

Williams, J. M. (1976). Synaesthetic adjectives: A possible law of semantic change. *Language, 52,* 461-478.

Worchel, P., & Dallenbach, K. M. (1947). "Facial vision": Perception of obstacles by the deaf-blind. *American Journal of Psychology, 60,* 502-553.

Zigler, M. J. (1926). A review of David Katz's Der Aufbau der Tastwelt. *Psychological Bulletin, 23,* 326-336.

Author's Preface

I entered the first notes concerning the world of touch in my scientific journal prior to the year (1911) in which my work on the World of Color (Katz, 1911, 1935) was published. My first thorough-going experiments on touch phenomena were carried out in 1914, at the Goettingen Psychological Institute. Even if the work could not be continued in laboratory fashion during the [First World] War, I still carried my most important instruments—my fingers—around with me, and thus was able to collect new empirical data uninterruptedly on the nearly inexhaustible realm of the world of touch. In the summer of 1919, I resumed the experiments at Goettingen, and completed them in the new Psychological Institute at Rostock, after moving there. A few of my results have already been published in two brief preliminary reports. Since both of these appeared in rather inaccessible places, I have included in the present monograph everything from them that seemed important. Thus, it will not be necessary to refer the reader back to those two reports.

Although my research should primarily interest psychologists, I hope that it also will be of use to workers in other disciplines as well, and will find application to everyday life in one form or another. In this respect, medicine comes first to mind. Investigations of the more complex tactual phenomena should nicely complement the studies of more elementary processes by sensory physiologists. Neurologists and otologists perhaps will be more interested in the material on the vibration sense, while general practitioners will be more concerned with the discussion of the psychological basis of palpation and the use of the hand in medical practice.

A leading proponent of the industrial school movement said in reviews of my two preliminary reports that he hoped my work would provide a psychological foundation for industrial and practical vocational instruction. I would be delighted if this expectation were even partially fulfilled. The remedial teacher, in instructing the blind and the deaf-mute, will be able to make use of most of the material in this book. Elsewhere in applied psychology, the human factors engineer who is concerned with measuring and improving the productivity of the human hand will not want to miss the present theoretical explorations.

Many psychology textbooks still divide the senses into upper and lower levels, with vision and audition ranked higher than touch. My research will show, I hope, that this lower valuation of the skin sense—which has prompted psychological researchers to neglect it as well—can in no way be justified, neither by the supposed poverty of phenomenological aspects nor by a second-rank standing in perceptual value. Indeed, we have good reason to believe that the current principles of the theory of perception would change if we deprived people of the skin sense, but not if we eliminated vision and audition. Thus, the theory of perception should stop relying so heavily on the higher senses as guides; much new knowledge and stimulation can be obtained by taking the lower senses into consideration. Schiller once said to Goethe, "It is high time that I close up my philosophical shop for a while; my heart yearns for something that can be touched."— I would like the following monograph to show that one can deal with tactual objects without thereby losing contact with philosophy.

David Katz

DIVISION I: MODES OF APPEARANCE OF THE WORLD OF TOUCH

DIVISION I: MODES OF APPEARANCE OF THE WORLD OF TOUCH 27

Section 1. Introduction

When I close my eyes and glide my hand over my writing paper, blotter, writing pad, and the cloth cover on my desk, I experience four distinctly different tactual impressions, one after the other. I feel the hard smoothness of the writing paper, the fibrous suppleness of the blotter, the leathery brittleness of the writing pad, and the soft roughness of the cloth cover. Once having become alert to these sorts of differences, if I now touch everything within reach, then I find that I not only can recognize the metal of the paper weight, the glass of the ink bottle, and the wood of the penholder, but that I can even distinguish—naturally also with my eyes closed—between different types of paper by their surface features, which not even the eye can do. We make a more or less conscious, but very full use of such tactual abilities in our daily life, as when we fish out one of the various types of paper in the dimly-lit desk drawer, or find the sheet we want in the dark linen closet, or when the experienced housewife, in shopping for some fabric, glides her scrutinizing hand over the samples to determine the quality and durability of the material and weave. When such tactual actions have been practiced for years or decades in one's profession, the skill achieved may seem miraculous to the lay person. Experienced fabric and rug dealers recognize, by touch, fabric blends (e.g., the slight presence of cotton in woollens) that would not be detectable by our hand or by our eyes or even by the dealer's own eyes. In large wool-producing countries, such as South Africa and Australia, people learn through continual practice to discriminate by touch the finest variations in the quality of samples of raw wool, and obtain considerable material rewards from this ability. Psychology has taken scant notice of these and similar holistic feats of the touch sense, and has not recognized what a rich profit it could obtain by explaining them.[1] In the introduction to the World of Color (Katz, 1911, 1935), I noted that many color phenomena once were regarded as marked curiosities, and as a result, useful connections with other facts of visual perception were overlooked. A similar situation occurs here. When one thinks of the literature on the tactual phenomena with which we are concerned here, then it is more apt to involve the appendix, where the rarities are found. However, these tactual phenomena definitely do not belong in the cabinet of psychological curiosities, where they may so easily be forgotten. Rather, we must pull them out into the daylight, dispel their characterization as sterile curiosities, and bring them into a more lawful association with other data on the touch sense. The rewards for doing this will be unexpectedly abundant, namely, the discovery of a surprisingly large realm of distinctive tactual phenomena, far surpassing in variety the realm of color phenomena.[2]

The World of Color (Katz, 1911, 1935) provides a how-to guide for discovering the relevant data. Striving for theoretical impartiality, we surveyed the whole set of tactual forms,[3] considering especially cases in which the commonplace character of the phenomenon threatened to obscure from our view what was baffling about it. However, it was not only the overall approach which I extended from the study of colors to that of touch. Even at the micro or detailed level, I followed the same path in investigating the phenomenological facts and the causal-genetic relationships of touch impressions as those of color. Extensive parallels were uncovered which, when traced back, revealed many new things in the color realm or shed new light on much that was already known.

Our experiments involved primarily, but not exclusively, tactual activities of the hand. The hand is a wondrous tool, whose power, in contrast to that of the highly specialized organs of other animals (grasping foot, swimming foot, climbing foot, running foot), lies in its astonishing versatility. Our studies will reveal the almost completely overlooked significance of the hand in the mental and intention-directed tactual world of the normal human. Using a slight variation of one of Schopenhauer's themes, one could speak of the world of touch as under the dominion and representation of the hand (see Section 12 below). Kant once used a very apt figure of speech when he called the hand, man's outer brain.[4] Lotze echoed this sentiment when he said "that a considerable part of human civilisation depends on the structure of the hand, and on the ease with which it enables us to make innumerable observations, to nearly all the lower animals rendered either impossible or but accidentally attainable by the comparatively imperfect formation of their organs."[5] Nearly thirty years ago, Féré quite rightly charged, considering the facts at that time, that psychologists had devoted too little attention to studying the hand. "The hand is simultaneously an agent and an interpreter in the development of the mind; it deserves more attention from physiologists and psychologists, who have somewhat neglected it."[6] According to Koehler's portrayals, apes are very close to humans in manual dexterity, as shown, for example, in grooming, in inspecting the skin, and above all in such activities as removing a small wooden sliver from a person's hand.[7] Did the human brain, in its ascent, create in the hand an organ commensurate with the creative ingenuity of the brain, or did the astonishingly fine sensory-motor coordination of the hand spur the brain to new capabilities on its part? To be sure, from a certain level of development onward, the relationship must be regarded as one of the most intimate interaction. Observations on children indicate that the demands of the brain spur the hand to ever greater improvement in its functioning. A child who has seen how adults snap on the electric lights, can have the strongest desire to imitate the adults, but fail to do so due to the clumsiness of the little hand.[8] When the effort finally succeeds, it is the brain which has educated the hand.[9] This case is paradigmatic for the entire development of manual dexterity in the child. The hand then repays its education by enabling the child to obtain the deepest and most practical insight into the structure of things and to penetrate into their innards in a completely different way than with the eye, which as a rule is confined to the outer surface of objects.

The [First World] War provided great impetus for research on the physiology and psychology of the hand, owing to the problem of amputees (Katz, 1921). Efforts to create an efficient artificial hand spotlighted the predominant position of this universal instrument and indicated very quickly the necessity of dispensing with a true copy and of obtaining prostheses capable of performing one or another partial function.[10] And even these prostheses lacked the sensitivity which animated the natural hand "up to the fingertips."

Recently, human factors psychologists have also turned to the study of the hand as well as to certain of its individual functions. In regards to the question cited above concerning the relationship of hand and brain, it is interesting that Moede found the correlation coefficient between general intelligence and manual dexterity to be .242,[11] though I prefer to think that in general the value is higher. Giese made a special study of the "working hand".[12]

In a discussion of the modes of tactual impressions, Scheibner[13] pointed out how very apropos and necessary the investigation of the facts treated here is for the vocational school and particularly for practical instruction. In Division IV, we briefly discuss applications of our research to pedagogy and psychological testing.

Footnotes

1. G. Martius characterized that psychology which is based on wholes as analytic, and that which is based on elements as synthetic. *Bericht ueber den 5. Kongress f. exp. Psychologie* (Report on the Fifth Congress for Experimental Psychology). Leipzig, 1912.
2. Not only in psychology has there been a tendency recently to vigorously investigate the lawfulness of just such forms, which seemed to lie beyond the pale of lawfulness and therefore were outside of what was considered in theory; one encounters this even more frequently in biology and medicine, which again shows the extent to which the same attitudes towards research take hold in the various provinces of science largely independent of the specific area. Phenomena such as, for example, incomplete development of certain eye muscles, striking asymmetries of the head, congenital torticollis (Caput opstipum), can no longer be dismissed, as before, as insignificant tricks of nature, but instead they are pursued, with an attempt made, for example, to determine whether they perhaps can aid in the study of heredity (familial research).
3. The expressions, tactual forms, tactual structures, and tactual shapes, are used in our studies in a purely descriptive manner; no commitment to any of the cur-

rently debated theories is thereby intended. I generally would have avoided these names if other, equally succinct and fit expressions had been available in German. As in the case of the modes of appearance of color, I intended the research to have more of a phenomenological and experimental thrust than a theoretical thrust. If theory is to help rather than hinder progress, then it must not be crystallized too soon. This holds for all the empirical sciences, but especially, as its history shows, for psychology.
4. Gerhart Hauptmann praises the human hand in his novel, *Die Insel der grossen Mutter* (The Island of the Great Mother), as follows: "The hand replaces all instruments, and through its congruence with the intellect, it vouchsafes the latter a universal mastery.—The social structure of Europe and its inherent morality should be illuminated by this new orb of thought. . . . It cannot be overestimated what would happen if the hand, now looked down upon, were elevated to the highest nobility."
5. H. Lotze, *Microcosmus: An essay concerning man and his relation to the world*. Translated by E. Hamilton and E. E. C. Jones. 2d. ed. New York: Scribner & Welford, 1887. Vol. I, Book V, Chap. II, p. 587. The tools humans create frequently are modelled on animal organs that are adapted to special tasks. The control of these tools by the hand ensures humans of an additional advantage over animals. (See G. Schlesinger, *Der Einfluss des Werkzeugs auf Leben und Kultur* (The influence of tools on life and civilization), Berlin: Zentralinstitut fuer Erziehung und Unterricht, 1917. M. Ettlinger provides a valuable review on the extent to which animals use tools: Ueber Werkzeuggebrauch bei Tieren (Animals' use of tools), *Philosoph. Jahrbuch, 37*, Fulda, 1924.
6. C. Féré, La main, la préhension et le toucher (The hand, prehension, and touch). *Revue philos., 41*, p. 636, 1896.
7. W. Koehler, Zur Psychologie des Schimpansen (Chimpanzee psychology). *Psychol. Forschung, 1*, p. 30 f., 1921.
8. For some further observations, see D. Katz and R. Katz, Kinderpsychologische Beobachtungen (Observations in child psychology), *Kindergarten, 63*, p. 106 f., 1922.
9. Similarly, adaptation to the prosthesis occurs through the determination of the amputee. What can be achieved under favorable conditions using a Sauerbruch artificial arm borders on the miraculous. See C. ten Horn, Spaetergebnisse bei Sauerbruchamputierten (Subsequent results in Sauerbruch amputees), *Muench. med. Wochenschr.*, No. 7, p. 233, 1922.
10. The most important material assembled on this up to 1919 can be found in the so-called prosthesis volume of *Zeitschr. f. orthopaed. Chirurgie*, Berlin, 1919.
11. Praktische Psychologie, *2*, p. 303, 1921.
12. F. Giese, Zur Psychologie der Arbeitshand (The psychology of the toiling hand), *Bericht ueber den 7. Kongress f. exp. Psychologie* (Report on the Seventh Congress for Experimental Psychology), p. 116 f., Jena, 1922.
13. *Arbeitsschule, 35*, No. 718.

Editor's Notes on Chapter I, Division I: Critical and Methodological Preamble (Sections 2 to 4)

In the next chapter, Katz attacks the atomistic viewpoint of sensory psychology (Section 2), describes more complex or holistic tactual phenomena which need to be explained (Section 3), and draws parallels between vision and touch that may elucidate the nature of the tactual phenomena (Section 4). He anticipated several of Gibson's (1966, 1979) concerns; he noted that rarely, if ever, in everyday life is only a single pressure or temperature point stimulated (Section 2), and he said that complex impressions like wetness and dryness are not inferentially constructed or obtained by summation, as Titchener proposed, from the elemental sensations of pressure and temperature (Section 3), but depend, like the impression of illumination in vision (Section 4), on simultaneous holistic processes. Although he rejected Helmholtz's model of unconscious inference, stating that complex tactual phenomena are not "cognitive products of logical operations," he provided no alternative computational account of how the sensory elements are formed into a percept, and thus he seems to offer an early version of Gibson's noncomputational model of direct perception. His focus, clearly, in on the phenomenology, not the mechanism.

Unlike Gibson, however, Katz did not depict the perceiver as resonating exclusively to invariants in the external stimulus. Katz provided a role for mental set as well (Section 4). Whereas visual objects are always projected "out there" (objective mode), a tactual object may be perceived instead at the sensory surface (subjective mode), and mental set can help to determine which mode (objective, subjective) predominates in touch. Katz did not say why mental set works, but presumably it is because the object touched is weakly specified or ambiguous in the sensory input. Katz noted that nothing in touch corresponds to illumination in vision, and that vision typically excels in discerning macrostructure (form), whereas touch is at least the equal of vision in discerning microstructure (texture). Recent work indicates that touch does indeed dominate vision in judgments of roughness (Heller, 1982, 1989; Lederman, Thorne, & Jones, 1986).

Chapter I:
Critical and Methodological Preamble

Section 2. The Atomistic Viewpoint of Sensory Psychology

As is well known, sensory psychology has developed in the closest association with sensory physiology. Vestiges of this path of development are plain for anyone to see. However, the gratitude owed by the daughter science ought not to blind it to the fact that it inherited a rather narrow-minded outlook from sensory physiology, which time and again has proven to be a definite impediment. As a result of this, the study of the so-called more complex or holistic phenomena has been neglected. To be sure, they have not been completely ignored, but they have not been accorded the central position which their importance warrants, and their true nature has been completely misunderstood in that they have usually been interpreted as cognitive products of logical operations.

Let us examine this in somewhat more detail for the skin sense, which is the focus of the present investigation. It is amazing how many excellent psychological observations E. H. Weber made in his classic studies on the sense of touch and common sensibility,[1] for example, his treatments of eccentric or external projection, objectification and subjectification of tactual impressions, the recognition of materials by touch, and the role which movement plays in this. It may sound paradoxical, but very likely we owe these unbiased and apt descriptions in part to the lack of knowledge in Weber's time concerning the sensory organs of the skin. After the epoch-making discovery by Blix, confirmed by Goldscheider and Donaldson, of the warm, cold, and pressure spots, a markedly atomistic approach took hold, directed towards the "elements" of the skin senses. No one followed up on Weber's work directed toward complex, lifelike phenomena. The atomistic model of natural science, already ascendant in the psychology of vision, received new impetus from the discovery of skin spots. According to this model, the most important task for sensory psychology is determining what sensations are produced when sensory points are stimulated. With this the danger is great that anatomical and histological considerations may suggest psychological facts that are spurious. For touch, no one has pointed out the necessity, as Hering did so decidedly for sight, of cleanly separating physiological and psychological matters.[2]

Quite correctly, sensory physiology considers that one of its most important tasks is to reveal the elements of our sensory organs and their manner of functioning. With the dissection of the nervous organism, sensory physiology attains its farthest reach. However, differentiation necessarily must be complemented by integration. Functionally, the sensory elements obtained by dissection are, in the final analysis, artifacts. In the living organism (whose expressions, after all, are what we wish to understand), large coalitions of sensory elements always work together.[3] Although Hering championed quite early the notion of interaction between the elements of the visual field, the extension of this holistic approach to other sensory areas, particularly to the one that interests us here, occurred only later and hesitatingly.[4] The atomistic approach, which obviously meant more to the people of those times, predominated completely. The onesidedness of the research indisputably brought great successes and no immediate serious disadvantages for sensory physiology.[5] It had entirely different consequences for sensory psychology. As some examples given below will show, it was a momentous error to believe that, using a kind of psychical chemistry, one could erect sensory perception from the elementary sensory sensations and certain concepts. The true nature of the more complex phenomena was overlooked, and, as a result, a false theory was foisted on the field. It is no wonder that the controversy between nativism and empiricism did not come to an end.

The only natural starting-point for research by the sensory psychologist is the naturally-arising phenomena found in consciousness, whether or not they then prove under analysis to be elemental or complex.[6] The statement above that the onesidedness had no immediate serious disadvantages for sensory physiology actually holds true only so long as the tacit agreement on the division of labor, whereby physiology treats the elemental and psychology the complex, is adhered to. Whoever rejects this agreement will also demand that sensory physiology elucidate the synergy between sensory elements, that underlies complex phenomena.

Neither in psychology textbooks nor in monographs on the skin sense can one find particulars on such tactual activities as those which formed the starting-point for my introduction. To tie our own investigations to preceding ones for now, let us consider the treatment of the relevant senses presented by Titchener in his sensitively-tuned textbook of psychology, which is distinguished by its clarity.[7] Then, it is evident first of all that our tactual activities primarily involve Titchener's touch (pressure) sense, and also the temperature and kinesthetic (muscle, tendon, joint) senses. Pain, which together with pressure and temperature forms the triad of skin sensations according to Titchener, has no separate standing, but rather represents only a nuance or shading for the collection of activities which I examine. American psychologists treat these senses atomistically, as is typical with all other psychologists. The temperature sensation has only a "substance" character, it lacks all shape, and much the same is true of the pressure sense. Research was concerned first and foremost with the properties of the sensations produced by stimulating individual pressure points. Even if one included investigations on the spatial sense of the skin in which several pressure points were stimulated, that would not really change the atomistic character of the studies on the pressure sense. It now must be emphatically stated that all of the experiences touched on here are artifacts of the consciousness. Thus, just as the excitation of a single, isolated sensory organ, e.g., a pressure point, occurs in an artificial way, so, too, the resulting state of consciousness is also not a natural growth, but rather the product of a quite constrained and complicated process.[8] One has only to consider all of the instrumental measures needed, in order to precisely stimulate individual sensory spots, e.g., pressure, warm, and cold points. To produce such experimental conditions means devising situations so extraordinarily remote from natural stimulus conditions that even the absurd, multiform accidents of everyday life would hardly ever lead to such situations and thereby to a sharply isolated stimulation of a single sensory element. Most people may die without ever having experienced the triggering of an isolated pressure or warm point or a genuine two-point threshold.[9] As valuable as it is to know how our nervous system can be excited by artificial procedures, it was the need for elucidation of natural activities that originally led to these procedures. Certainly, as Wundt has said, the principle of heterogeneity of purpose in sensory physiology prompted investigations so remote from the base of natural activity that one no longer considers their actual origin, but frequently traces back from the artificially produced state of consciousness to that arising naturally.[10] This is less true for naturally occurring tactual phenomena, because these generally have received very little consideration, as will be shown in the following paragraphs by the compilation of the most important complex natural phenomena cited in the literature.

Footnotes

1. E. H. Weber, Tastsinn und Gemeingefuehl (The sense of touch and common sensibility or coenesthesia). Ostwald's *Klassiker der exakten Wissenschaften*, No. 149, Leipzig, 1905. This paper first appeared in R. Wagner's, ed., *Handwoerterbuch der Physiologie* (Pocket dictionary of physiology), Vol. III, pt. 2, pp. 481-588, Brunswick, 1846. For excerpts in English, see R. J. Herrnstein & E. G. Boring, eds., *A source book in the history of psychology*. Cambridge, Mass.: Harvard University Press, 1965, pp. 34-39.
2. Rarely does someone like O. Funke, Der Tastsinn und die Gemeingefuehle (The sense of touch and common sensibility or coenesthesia), warn against drawing conclusions about the identity of sensations from anatomical facts. Hermann's *Handbuch der Physiologie* (Handbook of physiology), Vol. 3, No. 2, p. 289, 1880.
3. C.S. Sherrington primarily emphasized integration in many of his works on the organism as a stimulus-response system. The anatomy text by H. Braus, Vol. 1, Berlin, 1921, signifies a radical break with the analytical method of descriptive anatomy, where the atomistic procedure naturally is most deeply rooted. "Where previous texts dealt with the anatomy of the corpse, this text deals with the anatomy of the living human. Where the previous texts proceeded from the end product of dissection, from dead bones or muscles, this text proceeds from the living whole" (C. Elze in a review of the book, *Naturwissenschaften*, No. 43, 1921).
4. M. von Frey, along with his pupils R. Pauli, K. Hansen, and others, was the first to thoroughly investigate the issue of interaction between pressure sensations (Vols. 56, 59, and 62, *Zeitschr. f. Biol.*, Sitz.-Ber. d. Phys.-Med. Ges. zu Wuerzburg, 1917). "The psychological effect of a stimulus depends on its combination with excitation in adjacent circuits, a process known in the literature as summation (E. H. Weber), immediate induction (Sherrington), or mutual reinforcement (von Frey)." (The study cited last, p. 5.)
5. Of incomparably greater practical importance was the atomism of Virchow's cellular pathology, which long dominated pathology. Significant as that was for its period, so is the changeover from the part to the whole, that is, to the human personage, quite clear in our time. "Only an individual can become ill, obviously, even when the individual is a protist, that is, a one-cell organism. However, the concept of disease therefore does not apply for the individual cell of the kidney or the heart muscle, whose change is caused by the illness of the whole." L. R. Grote, *Grundlagen Aerztlicher Betrachtung* (Fundamentals of medical observation), p. 14, Berlin, 1921. Thus, we observe, too, in medicine a turn towards the person; such a breakthrough in psychology was aided by W. Stern in his last works.
6. On this, see the valuable reflections of H. Hofmann on the concept of sensation. *Untersuchungen ueber den Empfindungsbegriff* (Investigation of the concept of sensation). Dissertation, Goettingen, 1912.
7. E. B. Titchener, *A textbook of psychology*, Sections 38-50, New York: Macmillan, 1910.
8. Thus, these conscious experiences would not be designated as biogenic nor as

biomorphic (artificially produced, but biogenically equivalent) psychic events in the sense of W. Baade, Ueber darstellende Psychologie (Representative psychology), *Bericht ueber den 6. Kongress f. exp. Psychologie* (Report on the Sixth Congress for Experimental Psychology), p. 29, Leipzig, 1914.
9. That we do not permit the sensory life of the newborn to begin with the sensory elements is self-evident from this. See K. Koffka, *Die Grundlagen der psychischen Entwicklung* (Fundamentals of mental development), p. 93 f., Osterwieck a. Harz, 1921.
10. What was said above actually holds true for all of psychology; two random examples illustrate this. First, as a rule, experiments in animal psychology have been suggested by perplexities or riddles in the natural behavior of animals and they have generally elucidated that behavior. However, animal psychology, through its purposive selection of experimental conditions, frequently makes the animal do things it definitely could not be assumed to do in its natural habitat. The second example is provided by child psychology. Artificial manipulation of the process of consciousness in the child is supposed to reveal its natural course to us. However, in many individuals it reveals instead completely new and unsuspected worlds; e.g., think of the eidetic studies by Jaensch and his students. E. R. Jaensch, *Ueber den Aufbau der Wahrnehmungswelt und ihre Struktur im Jugendalter* (On the perceptual world and its structure in children), Leipzig, 1923.

Section 3. The More Complex Tactual Phenomena in the Present Literature

In one chapter of his textbook cited above, Titchener (1910) describes "A few touch blends," and we will quote the most important passage here. "The difference between hard and soft . . . is mainly a difference in degree of resistance offered to the hand; and this means a difference in the degree of pressure exerted by the one articular surface upon the other. The distinction thus belongs to the joints rather than to the skin. Again, the difference between smooth and rough is a difference, first, between continuous and interrupted movement, and secondly between uniform and variable stimulation of the pressure spots of the skin. The distinction thus belongs to the joints and skin together (p. 171)." It should be briefly noted here that hard can be discriminated just as well from soft, and smooth from rough, with a moving material. Thus, the discriminations can be evoked in a different way as well. "Sharp and dull differ, primarily, as pain and pressure: a thing is sharp if it pricks or cuts, blunt if it sets up diffuse pressure sensations. . . . Wetness is a complex of pressure and temperature. It is possible, under experimental conditions, to evoke the perception of wetness from perfectly dry things,—flour, lycopodium powder, cotton wool, discs of metal; and it is possible, on the other hand, to wet the skin with water and to evoke the perception of a dry pressure or a dry temperature. . . . Clamminess is a mixture of cold and soft: the cold sensations and the pressure elements in the softness must be so distributed as to give the perception of moisture. . . . Oiliness is probably due to a certain combination of smoothness and resistance; movement seems to be necessary to its perception. Clinging, sticky feels may be obtained from dry cotton wool (p. 172)." What deserves emphasis in these statements is the fact that the impression of wetness, stickiness, and oiliness can occur without any stimulus elements being used which "in reality" are wet, sticky, or oily.[1] We will return later (Section 29) to these and other quite similar phenomena, which we term tactual equivalents. However, it is just as impermissible to attempt to derive the latter impressions from summation of the specified individual sensations as to derive the impression of illumination from the sum of excitation of retinal elements.

Ebbinghaus' treatment of the more complex tactual phenomena is much briefer.[2] "The impression of points consists of the sensation of a spatially-delimited pressure. For smoothness and roughness, hardness and softness, different types of pressure sensations combine with sensations of movement and resistance. Wetness is then mostly a combination of smoothness and coolness, whereas dryness as a rule contains something of roughness, and so forth." In other psychology texts and handbooks, our subject receives hardly more space, if it is mentioned at all.

Even when we turn to an extended discussion or monograph on pressure sensations, such as T. Thunberg's,[3] we find no fuller a treatment of the "composite skin sensations." "Wetness and dryness are not elemental sensations, but inferences built up in various ways." Here we encounter the viewpoint, openly expressed, which we have already rejected, namely that complex impressions like wetness and dryness are inferentially constructed. Further, if the congenitally blind also have these impressions, then visual sensations cannot be indispensable components of them. "Compound sensations involving the skin are the sensations of smoothness and roughness. A discussion of the conditions for the arousal of these sensations shows what exactly the components are." "Simply touching an object, even with a very sensitive part of the skin, e.g., a fingertip, provides no notion of the degree of smoothness or roughness of its surface. When the objects merely touch, or are touched by, the skin, only very gross differences can be observed between objects whose surface qualities differ considerably (e.g., sandpaper, various fabrics, paper of differing roughness, sheetmetal, etc.). Only if the fingertip is moved over the object touched, or if the latter is displaced over the stationary fingertip, are the conditions favorable for a true comprehension of the nature of the surface. Thus, the sensation is composed of the sensation of steady contact and the sensation of a simultaneous and easy movement of the touching surface with respect to an object, that is, in part from a skin sensation and in part from a muscle sensation. The steadier the contact sensation and the easier the movement of the touching surface, the greater we reckon the smoothness of the object to be." We will find that what Thunberg thus adduced about movement as a creative force is confirmed quite comprehensively by our own experiments. However, this will not provide an occasion for us likewise to explore the pathways of psychical chemistry taken by this eminent scholar in the study of the skin sense. Alrutz revealed his own doubts about taking such a path in the case of pressure and smoothness sensations, when he commented as follows in his noteworthy research on the impression of smoothness: "I cannot conceive that the sensation of smoothness is to be interpreted simply as the summation of successive, homogeneous pressure sensations. Therefore, I think that something else also plays a role in this."[4]

Our selection of examples should be sufficient to confirm the contention that the literature has hitherto given scant attention to the abundance of naturally-occurring complex tactual phenomena. It should be added that the picture essentially does not change if one also includes the excellent new reviews of the methodology for testing the skin senses and the organs of movement, such as those by M. von Frey[5] and K. Goldstein.[6]

Footnotes

1. The recent American literature more frequently presents experiments belonging here, e.g., oiliness is treated by L. W. Cobbey and A. H. Sullivan, An experimental study of the perception of oiliness, *American Journal of Psychology, 33*, 1922, pp. 121-127.
2. H. Ebbinghaus, *Grundzuege der Psychologie* (Foundations of psychology), 4th ed., edited by K. Buehler, p. 368, Leipzig, 1919.
3. In W. Nagel's *Handbuch der Physiologie des Menschen* (Handbook of human physiology), Vol. 3, p. 706 f., Brunswick, 1905.
4. S. Alrutz, Neue Untersuchungen ueber Hautsinnesempfindungen (New studies on the sensitivity of the skin senses), *Bericht ueber den 1. Kongress f. exp. Psychologie* (Report on the First Congress for Experimental Psychology), p. 44, Leipzig, 1904.
5. L. Tigerstedt, *Handbuch der physiologischen Methodik* (Handbook of physiological methodology). Leipzig, 1914.
6. E. Abderhalden, *Handbuch der biologischen Arbeitsmethoden* (Handbook of biological methodology), Abt. 6, Teil A (Vol. 6, Part A). Berlin and Vienna, 1923.

Section 4.   Visual and Tactual Perception

1. *Demarcation of the activities.* Before focusing on the system of tactual phenomena whose structural principle is derived from that of color phenomena, let us briefly examine the general relationship between visual

and tactual perception. Our eyes provide us with knowledge of the chromatic and achromatic colors in the full multiplicity of their modes of appearance. There apparently is nothing equivalent to color *matter* (black, white, red, green, etc.) in tactual perception. As to the color modes, the primary colors (surface, film, volume, etc.) were distinguished from the secondary colors, which are, in all their variations, creations of illumination.[1]

Nothing in touch corresponds to illumination, so there are no secondary modes of appearance as were distinguished for color impressions.[2] Frequently, even the primary spatial modes of appearance for color contain important optical clues to the object's physical properties, as, for example, in the case of transparent and translucent objects. Later, we will take note of corresponding modes of appearance in touch. More important information on the physical properties of things is provided by the visually-perceptible surface texture than by the color modes. In the perception of texture, touch is at least on a par with vision. Using a word aptly coined by K. Buehler[3] for phenomena which probably were first demonstrated in my color modes, we will speak of texture as involving visual or tactual *microstructure*. *Macrostructure* will refer to the larger geometric forms, whether two-dimensional or three-dimensional (triangle, circle, cube, cylinder). Figures lying in a plane (drawings) can only be seen, whereas bodies which occupy space can also be touched. Stereognostic experiments have been carried out in great numbers in normal, mentally-disturbed, and blind subjects, but even where the tactual proficiency in perceiving macrostructure is pushed to its utmost, as with the blind, it is not on a par with that of the eye—in contrast to the case in perceiving microstructure.

2. *Objectification and subjectification in color and tactual phenomena: bipolarity of tactual phenomena.* Sometimes the senses are divided into the near senses, in which the objects perceived are in direct contact with the sensitive portions of the body, and distant senses, in which the objects perceived stimulate the sensory organs by means of a medium (air, light). With this division, vision would be counted as a distant sense and touch as a near sense. Certainly some tactual phenomena, such as those involving the probe principle (Section 25), seem to contradict the latter assignment, but nevertheless it is generally true that touching objects requires direct contact with portions of our body surface, while the eye can obtain information on objects at "astronomical" distances. This contrast results in a remarkable difference in the essence of tactual and color impressions, which has already been touched upon briefly.

"It is not consistent with the original meaning of the word 'sensation' to call colors sensations. It is appropriate to that meaning to say that one senses pain, pleasure, warmth, cold, but not to say that one senses white, red, or black. Sensations mean in our language something that one perceives in or on one's body, but colors always appear outside of our body and especially outside of our eyes."[4] Color phenomena are always characterized by objectification; they are always projected into external space. This applies even for phenomena (e.g., after-images, intrinsic visual gray) that depend solely upon the state of the retina or the visual cortex, and not upon external objects. If such phenomena are termed subjective visual sensations, that only is because of the absence of an objective stimulus source, not because of an inherent property that permits them to be experienced as belonging phenomenally to our own bodies. Experiencing a color phenomenon as subjective always occurs on the basis of certain secondary cues. No matter how deeply our attention penetrates into a subjective phenomenon like intrinsic visual gray or dark light, it never gives even a glimmer of what we readily know to be a state of our own bodily self. It is quite different with tactual phenomena. A subjective component that refers to the body seems inescapably linked with a second component that refers to the properties of objects. We therefore describe tactual phenomena as *bipolar*. A light, tickling touch with a feather on a spot which, like the back of the hand, is seldom used for touching, can indeed closely approximate a purely subjective tactual sensation, but cannot completely hush-up the evidence as to the causative stimulus. On the other hand, there are tactual phenomena that under one mental set seem exclusively to suggest something objective, but under another mental set, quite unlike the case with subjective visual phenomena, permit us to attend to the sensation as such (by which, a state of our body is understood). The latter is a clearly given property, not merely one that has been inferred, and it can be localized spatially with reference to our body. At any moment, either the subjective or objective side of tactual perception may be dominant, but this bipolarity nevertheless persists.

According to the above, on the issue of subjectification and objectification, we can ask, firstly, just as for color: When do we apprehend a tactual phenomenon as such in which there is no objective stimulus, and only an endogenous excitation of the touch organs?[5] Secondly, we can ask why an objective stimulus sometimes produces a touch that is almost solely an intimation of something objective, and other times produces a touch having the character of a definite sensation.

Whether the objective or subjective pole predominates in a touch depends chiefly on what spot on the body is contacted. The subjective pole is more salient on spots used less frequently to identify things, e.g., inner ear, inner nose. Contact a part normally covered by clothing (back, chest, stomach), then the "contact sensation" predominates. The objective pole is more salient with more energetic movement than with gentle movement accompanied by tickling. Movement of the touching organ also favors objectification. Weber (op. cit., p. 111) noted that when one part of the body touches another part, it feels the stationary part as an object. A peculiar dissociation arises when a warm hand touches a cold forehead; the warmth of the hand is sensed first by the cold forehead, but, the forehead is felt as an object touched by the hand (Weber). Analytically directed attention makes the subjective more salient, but in everyday life, we are tuned differently. W. Schapp indicated something of the kind: "One must indeed distinguish two types of attitudes toward objects: The attitude of the practical person and that of the theorist concerned with knowledge. It is quite different to touch something than to lay my hand on it. Coming into contact with something and perceiving it are not variations on the same theme, but produce a jolt when one goes from one to the other."[6] One can certainly say with H. Rupp,[7] that "hardness, weight, form, size, and sharpness are projected to the object; we perceive them in the object, not on our body," but even so the subjective pole is not missing either in our experience of hardness, etc. Friedlaender's[8] experiments also may be mentioned here. He threw new light on the old theme of lifted weight by allowing alternation between the two poles that determine how the impression of weight is experienced.

Compared with touch, temperature impressions generally show a more decidedly subjective flavor; the objective pole may vanish completely when warm and cold are experienced as pure states of our body.[9] Finally, pain sensations on the skin generally occur exclusively in subjective guise.[10]

Footnotes

1. The impression of illumination does not arise by way of summation (this word is used in the sense of the Gestalt theorists) from the individual colors filling the visual field. In contrast to these colors, it represents, as I already showed in Chapter 1 of the World of Color (Katz, 1911, 1935), a unique phenomenon dependent on simultaneous holistic processes. By no means do I wish to apply the term "gestalt" to the impression of illumination.
2. One could think of touched surfaces in which the tactual phenomena (e.g., sticky, oily, wet) correspond to the secondary color modes. In the present study, we will dispense with the distinction of primary and secondary touch modes in the sense in which it was previously used (Katz, 1920, p. 6).
3. K. Buehler, *Handbuch der Psychologie*. 1. Teil: Die Struktur der Wahrnehmungen. 1. Heft: Die Erscheinungsweisen der Farben (Handbook of psychology. Part 1: The structure of perception. No. 1: The modes of appearance for color), p. 64, Jena, 1922.
4. E. Hering, Zur Lehre vom Lichtsinn. Sonderabdruck aus Graefe-Saemisch's *Handbuch der gesamten Augenheilkunde.* (Outlines of a theory of the light sense. Separate reprint from Graefe-Saemisch's Handbook of general ophthalmology), Teil 1, Kap. 12 (Part 1, Chapter 12), p. 4, Leipzig, 1905. Translated into English by L. M. Hurvich & D. Jameson; Cambridge, Mass.: Harvard University Press, 1964, p. 2.
5. Using the procedures which C. W. Perky and O. Kuelpe adopted in visual perception, objectification and subjectification can naturally also be studied for

tactual perception. See the discussion below (Addendum, Section 12) on eidetic tactual images.
6. *Beitraege zur Phaenomenologie der Wahrnehmung* (Contributions to the phenomenology of perception). Dissertation, Goettingen, 1910, pp. 33-34.
7. *Probleme und Apparate zur Experimentellen Paedagogik und Jugendpsychologie* (Problems and apparatus for experimental pedagogy and child psychology), p. 159, Leipzig, 1919.
8. H. Friedlaender, Die Wahrnehmung der Schwere (Perception of weight), *Zeitschr. f. Psychol.*, *83*, 1919.
9. Language also clearly permits recognition of the different nuances for the subjective and objective when the sensation is made the object of a judgment. "With temperature, we formulate our judgment either like that with vision (it is warm, it is cold) or in a way that would not be possible with vision (I am or I feel warm or cold)." U. Ebbecke, Ueber die Temperaturempfindung in ihrer Abhaengigkeit von der Hautdurchblutung und von den Reflexzentren (Temperature sensation as a function of dermal circulation and of the reflex centers), *Pfluegers Archiv*, *169*, 1917, p. 396. —The diversity of linguistic expressions may also occasionally feign an abundance of sensory elements, where in truth only the different forms of an elementary type of excitation are present. Excitation of the touch sense permits one to speak not only of touch sensations, but also of pressure, stretch, tension, tickle, resistance, and still other sensations.
10. J. D. Achelis recently strongly emphasized this: "Pain is *never* the property of an object, no more than being satiated is a property of food." "The pain sensation therefore is never objectivized, for it cannot be by its very nature." *Zeitschr. f. Sinnesphysiol.*, *56*, 1924, p. 38.

Editor's Notes on Chapter II, Division I: Types of Tactual Phenomena (Sections 5 to 13)

In the next chapter, Katz again asserts the need to study complex tactual phenomena (Section 13), claiming that there is nothing inferential or constructed in them. (His proof of the latter, though, is simply to ask the reader to feel the paper of this book and to decide for himself or herself whether anything inferential or cold-blooded is present in the percept.) Due in part to Katz's influence, current work has begun to focus on haptic object recognition and "the intelligent hand," with the emphasis extending far beyond the elemental sensations of Titchener and the Structural school (Klatzky & Lederman, 1987; Lederman & Klatzky, 1987).

Katz also covers tactual mental images, and the role that memory touch (an analog of Hering's memory color) has in mental imagery and in perception (Section 12). (To obtain a "tactual after-image," Katz advises, move your hand over a material and then suddenly stop. The impression of the material may persist while the hand is at rest, even though the impression would never have arisen if the hand had been stationary in the first place and had never moved.) His main concern in this chapter, however, is in the three modes of appearance of touch: surface touch, immersed touch, and volume touch. Volume touch is covered in Section 7 (also see Section 32) and immersed touch in Section 6. Immersed touch is the "official" tactual analog of film color, but two other phenomena are mentioned that also bear some resemblance to film color: 1) touch-transparent film, e.g., the veiling layer of a loose glove used in touching (Section 8), and 2) the wetness, oiliness, or stickiness through which one feels the grain of a piece of wood (Section 9), just as one sees a wall through the membrane of a shadow. Touch-transparent film will be covered again below in Section 31.

Surface touch is covered extensively (Sections 6, 9, 10, 11), with special emphasis on tactual texture (Section 10), and tactual continuity and the figure-ground relationship (Section 11). In Section 9, Katz distinguishes between the qualities (*Modifikationen*) of surface touch (covered again below in Sections 18 to 28), and the identifying characteristics (*Spezifikationen*) of surface touch (covered again below in Sections 29 and 30). Qualities involve dimensions such as rough-smooth and hard-soft, whereas identifying characteristics are those features which specify exactly what the material is, e.g., paper, silk, wood, etc.

Katz frequently compares touch with vision on the various phenomena. Thus, he notes the lack of anything like hue and saturation in touch (Section 5), the presence of analogous modes of appearance of touch and color (Section 6 and elsewhere), the great importance of texture for both vision and touch (Section 10), and the greater reversibility of the figure-ground relationship in vision than touch (Section 11). He also notes the close analogy between memory color and memory touch, but points out the greater subjectivity of the tactual mental image, which always bears a trace of the touching agent (Section 12).

Chapter II:
Types of Tactual Phenomena

Section 5. The Monotony of Tactual Matter
and the Polymorphism of its Modes of Appearance

1. *Putting aside the genetic viewpoint.* We first apply to the modes of appearance in the world of touch the method of simple description, which made possible a first-order overview of the color modes as well. Without prejudice to the later investigation of the "whence," we first ask only about the "how" of the world of touch. Let us first use the word *touch* in the most comprehensive manner of common-sense psychology,[1] without going into the question as to which sensory organs of the skin besides those of the tactual (pressure) sense proper, or what other organs besides those of the skin, contribute to the construction of the world of touch. Thus, we will disregard for now all physiological considerations, as well as any distinctions between the tactual experience provided by very sensitive areas, such as the fingertips, and that provided by less sensitive areas, such as the back. Just as we disregard the subjective conditions of the site of the stimulus, so we also disregard their objective conditions, the nature of the stimuli. One may deduce just from the context in each case whether the word "touch" is meant in a narrower sense than that of popular usage.

2. *Touch value or matter and the mode of appearance of tactual phenomena.* I briefly remind the reader that color value or matter is that which remains constant when the color mode varies. For example, basic red can occur as a film color, surface color, etc. What in the tactual realm corresponds to color matter, what to color mode? To be consistent with the color realm, when hardness, softness, roughness, smoothness, etc. vary, one ought to regard what remains constant in tactual experience as tactual matter, and what varies structurally as tactual mode. Every single discriminable color or achromatic tone may become the color matter of a color impression, that is, any element of the entire range of this three-dimensional [i.e., hue, saturation, brightness] variety. What corresponds to this on the part of the touch sense? Let us first quote Titchener, who, in his 1910 textbook (op. cit.), depicted tactual sensations briefly, but appropriately: "If with the point of a pencil you brush one of the hairs that are sparsely scattered over the back of the hand, you obtain a weak sensation, of bright quality, which is somewhat ticklish, and which though thin and wiry yet has a definite body. This sensation, which we may term the sensation of contact, is physiologically a weak

sensation of pressure (p. 146)." Titchener then describes how one can use a horse's hair in the well-known manner to obtain pressure sensations of varying intensity. "By applying the horse-hair to the pressure spot, with different degrees of pressure, it is possible to call out the pressure sensations at different degrees of intensity. You get, first of all, the wiry, bright sensation of the former experiment. As the pressure is increased, the sensation too becomes heavier, more solid: at times it has about it something springy, tremulous, elastic; at times it appears simply as a little cylinder of compact pressure. Finally, at still higher intensities, the sensation becomes granular: it is as if you were pressing upon a small hard seed embedded in the substance of the skin. The granular sensation is often tinged with a faint ache, due to the admixture of a pain sensation; and is sometimes attended by a dull, diffuse sensation derived from the subcutaneous tissues. It may, however, appear as pure pressure sensation (pp. 146-147)."

Since the preceding description, of all those known to me, grants the richest endowment to tactual (pressure) sensations, then we may indeed say that the latter differ from color sensations in having a quite unusual monotony. Nothing in touch corresponds to the difference between chromatic and achromatic stimuli, nothing to the different qualities of the color circle. If one grants that there must be an intensity dimension for achromatic stimuli, then there would indeed be a corresponding dimension of pressure intensities. But the set of discriminable intensities and other qualities (*Modifikationen*) still to be considered for the tactual sensations (see Titchener's statements immediately below) seems quite modest compared to the plentitude of sensations shown by the black-white dimension alone.

This comparison of color and tactual matter points up, quite significantly the *extraordinary monotony of touch matter. The polymorphism of the world of touch contrasts strikingly with the monotony of its matter.* The touch matter is molded into a world of forms at least as rich and varied as the world of color. This implicates strong, diverse central processes in the erection of the world of touch.

The contrast which we have termed "monotony of tactual matter vs. polymorphism of the world of touch," led Titchener in 1910 (op. cit.) to state the following, which I now reproduce in full due to its relevance for later observations: "It seems, at first, hardly credible that the end-organs of pressure should not be differentiated for the reception of different kinds of stimuli. When we think of the great variety of our tactual experience, and when we remember further that the same stimulus has markedly different effects if applied to different parts of the skin, we are almost forced to believe in a number of qualitatively distinct sensations. Nevertheless, the verdict of experiment is decisive here. . . . And we must not forget the facts on the other side. First, the stimuli that normally affect the skin are areal stimuli, appealing to a group of diversely tuned pressure organs; and the texture of the skin itself, and the nature of the underlying tissues, vary from place to place. There is, then, every chance in ordinary experience for typical differences in the intensity and the temporal course of pressure sensations.[2] Now the sensations which we have termed contact, pressure, and granular pressure, although they are evoked by different intensities of the same stimulus and are on that account usually considered as different intensities of the same quality, are at least as distinct as red and pink, or yellow and orange; and if we may not call them psychological qualities, we must at least say that they do the same service for touch that true qualitative differentiation does for other senses. Secondly, the greater number of normal stimuli affect other organs, cutaneous or subcutaneous, besides those of pressure. Hence most of our tactual experience does, in strictness, consist of more than one quality, because it derives from more than one sense. Thirdly, as has been said above, the attention is generally concerned rather with the stimulating object than with the sensation which it excites. Here touch borrows from sight in much the same way as taste borrows from smell; visual characters of form, size, texture, etc., are so firmly associated to the feel of the stimulus that the skin gets the credit of a good deal of work done by the eye (pp. 148-149)." Disregarding details for now, Titchener's statement confirms our view as to the significance of central psychological processes in the erection of the world of touch.

The comparison of color and tactual phenomena makes it appropriate to briefly consider a technical matter regarding the different ways they are produced. The methodology of color psychology has been well developed for a long time; the visual spectrum offers a clear means of reproducing identical colors. Procedures for producing the particular modes of appearance are also simple, clear, and easy to describe. In touch, there is no such means of representation as the spectrum. The apparatus for producing various tactual forms is certainly not complicated, but it nevertheless is less clear and requires a detailed description in each case.

Footnotes

1. According to Grimm's dictionary, the German word for touch, "tasten," comes from the Italian "tastare," which derives from a Middle Italian "taxitare," the iterative form of Latin "taxare," to touch or handle.
2. A brain-damaged individual suffering a complete loss of visual imagery could not describe differences in quality due to contact made at various points on the skin. According to Goldstein and Gelb, this argues against the supposition of some researchers "that every discriminable point on the skin has a particular qualitative coloring." K. Goldstein and A. Gelb, Ueber den Einfluss des vollstaendigen Verlustes des optischen Vorstellungsvermoegens auf das taktile

Erkennen (Effect of complete loss of visual imagery on tactual recognition). *Zeitschr. f. Psychol.*, *83*, 1919, p. 12.

Section 6. Surface Touch and Immersed Touch

The first distinction to arise from our analysis of color phenomena was that between film and surface color. It is easy to demonstrate that there are tactual phenomena which correspond in almost all respects to surface color, and by analogy we will designate them as surface touch. We experience surface touch when we feel and manipulate an object made of wood, metal, glass, cloth, or other material. What is common to all of these tactual phenomena? In each case, we encounter a continuous, unbroken palpable area, which is located at the surface of and follows all curves on the object in which it occurs. It makes no difference whether we have a completely unyielding material before us, such as glass, or whether we touch soft woollen material spread out on a hard support. Invariably, a two-dimensional tactual structure, an obstacle bounded in space, presents itself to our consciousness. A surface touch, like a surface color, can be experienced at any particular location in space. Its orientation and distance from the perceiver can vary greatly from case to case (naturally, always within the limits set by our physical body), but are solidly fixed in each particular case. Just as the surface color points to the visual properties of an object, so the surface touch provides us with the tactual properties—though the two phenomena do not work in completely the same way, as later discussion will show.

When we spoke above of the localization of surface touch in space, we meant the tactual space of the normal sighted person with the eyes closed. We also presuppose normal observers in all of the following analyses. For now, we will not discuss the nature of the relationships existing between visual and tactual space. In Section 46, we will consider in detail what influence vision has on tactual perception.

Are there tactual phenomena comparable to film color? One might suppose that these could be obtained by making the surface touch fainter and fainter, as in going from hard materials, such as wood and metal, to increasingly softer materials, such as cloth and cotton. But even the softest and faintest surface always provides only tactual impressions which, by their structure, are surface touches. Tactual impressions corresponding to film color occur instead in phenomena of the following type. Direct a powerful stream of air against the hand, or move the hand with sufficient speed in liquids of various consistency—then one experiences a tactual phenomenon that has no definite shape or pattern. It has a certain thickness, but cannot confidently be regarded as spatial, because it lacks the rear boundary found in volume touch, which we will go into below. Also, unlike surface touch, our tactual phenomenon does not have a fixed orientation in space. Lacking a better term, the tactual phenomenon corresponding to film color will be called *immersed touch*. The new tactual phenomenon does not stand at quite the same level as film color in regards to its function in perception. Nevertheless, like film color, it is characterized negatively in that it does not represent a property that would make the object confidently recognizable. Immersed or space-filling touch can characterize only a substance and not an object. If an immersed touch is obtained through one of the means mentioned, then the flimsiness of its structure, which nevertheless does not take on a pronouncedly fluctuating or interrupted quality, may be experienced. (I must repeat here what I occasionally stated in The World of Color, Katz, 1911, 1935, that even the most detailed portrayal cannot provide a completely satisfying picture of such elementary phenomena.) The subjective pole emerges more markedly in immersed touch than in surface touch, but it would be wrong to regard the phenomenon as exclusively subjective. The consciousness of something palpable and objective is present, even if this also bears rather the character of a momentary creation and the mark of transitoriness.[1]

As stated previously, a surface color offers resistance to the gaze, whereas in a certain sense we can penetrate into a film color. Both statements naturally were meant figuratively, since genuine experiences of resistance are encountered only in touch. What about surface touch? An experience of resistance *can* occur in conjunction with it, but does not *have* to, so far as I can judge. If one touches some surface lightly with the fingers, for example, one of wood, then a definite surface touch occurs without an accompanying experience of pronounced resistance. If the pressure of the fingers is now increased on each new touch, up to the uppermost limit, then a feeling of resistance enters at a certain point and increases to a maximum. However, the surface touch remains essentially invariant throughout the entire interval from where the phenomenon of resistance just begins up to where it attains its highest degree of pronouncedness. If the experience of resistance belonged inseparably to surface touch, then surface touch would have to be altered completely by such an extensive variation of the experience of resistance.[2] Its property of *impenetrability* can very easily mislead one to consider the resistance experience to be an inseparable component of surface touch. The inability to penetrate is actually an important property of the surface touch; this property changes into active resistance against the touching organ when the latter intensifies the pressure beyond a certain limit. The matter is entirely different with the tactual phenomenon we have called immersed touch. One cannot imagine removing the experience of resistance without thereby eliminating the sensation itself; it changes with the active force of the flow which strikes the hand, disappearing when this becomes zero. The resistance characterizing this experience is not stiff and rigid, but elastic. Whereas the film color is arrayed essentially in frontal-parallel fashion, the touch sensation here is oriented perpendicularly to the touch organ.

In contrast to vision and audition, the senses of the skin and the associated kinesthetic senses have been treated as stepchildren by the psychologist, quite unjustly, since, as shown above, they contain the psychological roots for the physical concepts of impenetrability, resistance, and strength. In view of these and other facts, we agree with J. Petzoldt that mechanics originated from the sense of touch just as optics originated from vision (*Annalen der Physik, II*, Leipzig, 1919).

Footnotes

1. The moving atmosphere strikes us subjectively as an immersed touch. This may well have prompted [so-called] primitive people to make animistic interpretations of inanimate natural processes.
2. According to F. Kiesow (*Arch. f. d. ges. Psychol.*, 22, 1911, p. 62), H. Aubert and O. Kammler (*Moleschotts Untersuchungen, V*, 1858) also affirmed that there is a "pure touch sensation" and "touch without pressure." The sensation of touch or contact was supposed to differ from that of pressure, but be evoked nevertheless on all positions of the body by a physical pressure that varied only in intensity.

Section 7. Volume Touch

Place a small object, such as a matchbox, on a solid support, and cover it with a thick layer of cotton wadding or cloths. If the object is then felt in order to recognize its form, we obtain a pretty good idea of what it is, and at the same time the filling material lying above it provides us with a space-like or volume touch. In trying to recognize the object underneath the filling material, one is no longer aware of the surface of the filling material. This surface seems to have become transparent, and the impression the material evokes is completely that of a space filled with a soft mass.

Colors seem to be genuinely voluminous only when objects are seen through them. Volume colors are seen when a vat filled with a milk solution is placed in front of objects, but not when it is placed in front of the sky with its film color. Likewise, a layer of cotton wadding does not produce a definite voluminous tactual experience as long as it rests on a level support and no sculptured object is felt through it. The wadding produces a surface touch to light pressure, and a soft, but not clearly a space-like or volume touch when the hand is moved with somewhat greater pressure. Just as volume color is most perceptible with a medium degree of opacity or cloudiness, so too is volume touch most perceptible with a medium degree of thickness of the layer of cotton. The soft mass of cotton seems to surround the solid object very tightly; it ends where the surface of the object appears to begin. Just as volume color has that much clearer an effect, the more impressive the bodily quality of the objects lying behind it, so too is volume touch the more pronounced, the more definite the physical objects lying under the layer of cotton. The volume color of fog swallows up details of the surface texture of the objects and is particularly definite at their edges; likewise, the cotton wadding also covers details of the surface and is pronouncedly voluminous only at the edges of the touched objects. If fog becomes too thick, then its voluminousness disappears along with the objects; the voluminousness of the tactual phenomenon is lost in precisely the same way with a very thick layer of cotton that does not permit the surrounded object to be felt.

In medical practice, volume touch plays a certain role, without, to be sure, receiving great notice as such. By palpation, the physician "touches" the internal organs through the skin and cushions of fat in order to detect pathological changes in them. The attention is directed at the organs themselves, and not on what lies between them and the feeling hand and is given as volume touch.

Section 8. Touch-Transparent Film

There also are tactual analogs of color transparency. Take a piece of paper or cloth that is not too thick, and, without letting it glide past the fingers very much, move it back and forth over a fixed, underlying base made of wood, metal, or any other material. One then experiences the underlying base, whose surface texture will be perceived with varying clarity, depending upon the thickness of the paper or cloth covering the fingers, a clarity which is surprisingly great considering the veiling effect of those intervening media. Now, while feeling the texture of the underlying base, one does not for a moment lose the impression of a thin film in front of the fingers through which the underlying base is touched. The experience of the touch-transparent film persists in all its clarity, along with that of surface touch, with no hint of a fusion of the two tactual impressions, so long as the motion continues. With appropriately selected experimental conditions (a rather light sliding of the intervening medium) and properly directed attention, one can simultaneously perceive the surface texture of the underlying base, the surface texture of the veiling layer, and the veiling layer itself.

These tactual experiences are common in everyday life as well, occurring every time one touches something while wearing a glove. In this case, the fabric or material of the glove represents the touch-transparent film. The physician encounters such tactual phenomena as well, as, for example, when he or she undertakes palpation while wearing rubber gloves to prevent infection. Here, to be sure, the touch-transparent film is less pronounced, because the membrane fits so tightly on the touching organ as to prevent even the slightest shifting, and because the membrane is generally so very thin. With very thin rubber gloves, the impression of merely having a certain numbness in the fingers can replace the impression of being covered.

Just as the boundary between transparent film color and volume color is a fluid one, so too, can all transitions between volume touch and the impression of the thinnest touch-transparent film be achieved by suitable experimental arrangements.

The importance of motion for the clarity of the touch-transparent film ought to be noted. I might add to my earlier statements concerning the transparent colors in the World of Color (Katz, 1911, p. 15 f.) that they, too, increase extraordinarily in clarity when the intervening medium used (smoked glass, gelatin, episcotister) is moved, or when the head is moved relative to the intervening medium. The increase in visual transparency becomes quite apparent when one moves a crepe paper lying over a printed background, though to be sure, physical factors become involved here along with the psychological ones of transparency.[1]

Footnote

1. W. Fuchs (1923) also recently pointed out how color transparency increases in clarity as a result of motion. *Zeitschr. f. Psychol.*, *91*, 1923, p. 220.

Section 9. Qualities (*Modifikationen*) and Identifying Characteristics (*Spezifikationen*) of Surface Touch

The dazzle of colors in the world proves under closer examination to be a profligacy that contributes less to visual orientation than one might have assumed a priori. Even totally color-blind persons, who see themselves as completely deprived of the aesthetic pleasures available to those who see colors, do not make their way about that much more poorly. They are sensitive to space and form, just like normal persons. In recognizing material, the microstructure sustains them. A piece of paper may have all of the colors available in the system of surface colors, but it still will be recognized as paper by its texture, which is accessible to the color-blind person. Color can deceive, but texture cannot do so as easily. Having previously pursued the matter of the visual surface structure (Katz, 1911, p. 91 ff.), the question now concerns the varieties of tactual surface structure.

The description above (Section 6) of surface touch is somewhat oversimplified, because in reality it would be just as unlikely for a surface touch to possess only the properties described there, and none of the other distinguishing properties, as for a visual surface to have no particular texture. We never have the impression of a "general" surface, but always that of a hard or soft, rough or smooth surface, etc. Just as all gradations occur from a hard metallic to a soft cloth surface, so, too, do we experience all the in-between steps from the greatest smoothness to the most pronounced roughness. It is not difficult to point out examples of particular combinations of properties from the hard-soft and smooth-rough dimensions. This is made clear with four extreme instances: glass has hard smoothness, sandpaper has hard roughness, silk has soft smoothness, and billiard-table cloth has soft roughness.

For ease of exposition, we will designate all of these particular features of a surface touch as its qualities (*Modifikationen*). A quality of a surface touch therefore signifies classification along such dimensions as hard-soft and rough-smooth; it refers to a general characteristic of the surface touched, but not of the material itself. It is striking how meager language is in providing separate designations to express the qualities of surface touch. For the rough-smooth dimension, as far as I can see, language provides only the basic expressions smooth, dull, and rough. In saying this, we naturally disregard the comparative forms of these basic expressions, such as very smooth, very rough, etc., as well as technical expressions, such as polished, scrubbed, knurled, etc. Even if psychological research seems to have shown that there are fewer values on the hardness dimension than the roughness dimension, the poverty of language is yet striking again in this case, with only two designations provided for the end members, namely, hard and soft. On the rough-smooth dimension, the number of discriminable values is hardly less than that of the black-white dimension. The linguistic designation of touch experience, rough-dull-smooth, can be set in parallel to that of color, black-gray-white.

To express something more definite about a tactual impression than is allowed by general linguistic expressions, we will resort to the same device used in designating colors for which there are no basic expressions. Just as we may speak of violet blue, ruby red, emerald green, or chocolate brown, so too, can we designate surface touch as being leathery, cloth-like, silky, or papery, in order to characterize it as similar or equivalent to the impression previously experienced when touching leather, cloth, silk, or paper.

We will designate as identifying characteristics (*Spezifikationen*) of surface touch those tactual experiences that refer to a certain fabric or a certain material. The number of discriminable identifying characteristics is extraordinarily great; it also depends upon the experience of the perceiver. Naturally, any identifying characteristic of surface touch can be "modified" or changed in quality in this or that direction; there is smooth and rough, hard and soft paper, wood, etc. If we recognize some suit material as *our* suit material, some wood as *our* work stool, then the most individualized form of specification has occurred, with a tactual impression being established as unambiguously as possible and recognized as such.

A wood surface can be smooth or rough, hard or soft, but impressions of a quite different type arise if the wood feels wet, oily, or sticky. Properties of this type occur on the surface touched in a way similar to that of a lit or shadowed spot on a color surface. One touches the specific qualities of the wood through the wetness, oiliness, or stickiness, but these latter qualities yet have some influence on the impression of woodiness. When we speak here of touching through, then this naturally has a quite different sense than when we spoke above of touch-transparent films (Section 8).

Section 10. Natural and Artificial Forms of Materials

1. *Visual texture.* All materials confront us in natural or artificial forms. Stone exists as naturally-formed rock, or in its artificial form, as the slab polished by human hands. We must confess that it is not always easy to determine whether something is in a natural or artificial form, but, as a rule, it is possible to decide. That decision is not unimportant for our orientation in the external world. Let us pursue this matter first in the visual realm.

If a wooden board is lying before me, I recognize the material immediately as wood in normal illumination. This recognition is quite independent of the artificial color given to the wood, as long as no coat of paint completely hides the grain. The board may gleam in all the colors of the rainbow, but nonetheless the texture of "wood" would be everywhere evident. Recognition of the material is also independent of the outer shape of the wood. It does not matter whether the board has large flat or curved surfaces, or whether the carpenter has given it an elaborate form. The texture of the material reveals itself at every point in the artificial form. The same would have been true if we had chosen a different material than wood. How is the texture revealed visually? Describing this matter is remarkably difficult, and I must ask the reader to aid me in the endeavor by carefully observing the material structure of the printed page lying before him or her. At a certain illumination and distance from the eye, which are quickly determined by trial and error, the texture appears most clearly. One then discovers formal elements, very small and hardly differentiated from each other, which owe their visibility to minimal differences in brightness and hue. These elements are so small that a great number of them probably could be discovered within only a square millimeter. There is an astonishing variety among these elements. When examined closely, no two will be found to be exactly alike, and yet, for all this irregularity, there is a certain regularity in the recurrence of the elements over the entire surface. There is, so to speak, a type of element which is the vehicle for the impression of "paper of this kind." We might even say that regularity within irregularity of elements is the law of texture. I have already indicated several times that the participation of the cones in texture perception has a much greater biological significance than their participation in color perception.[1] There are materials in which the smallest formal elements are combined into structures of higher order, and these, in turn, into structures of an even higher order, which then give the material its characteristic texture. In the case of wood, consider the transition from the smallest elements to the suggested grain, and from that to the distinct grain shown by the annual rings. Depending upon the circumstances in which the judgment is to be made, one attends to the elements or to higher structures of this or that order in orienting oneself towards the material. The structural elements can determine our judgment concerning the material without our actually becoming conscious of them. The color and brightness differences which constitute the textural elements of the paper have no effect at all on our judgment of the color of the paper. Following Hering, we may say that a colored paper mounted on glass has ideal uniformity of color, and yet even so it does not entirely lack those differences in color and brightness, for if it did it would no longer be seen as paper, but only as a film color.

When a certain artificial shape is impressed upon a material, then new structures of formal elements occur. These structures can be extraordinarily small; consider, for example, the finest cloth woven from silk. Yet they nevertheless always must be larger than the elements which are characteristic of the material itself (silk). However, the elements of natural and artificial forms differ not only in size, but also in regularity. The elements of the artificial forms exhibit a much greater regularity, and in the limit can all be exactly the same.

2. *Tactual texture.* Much more could be said from a phenomenological perspective about these things, which hitherto have almost completely eluded investigation, but we will break off at this point—even the previous statements concerning vision would not have been appropriate, had they not been valid in almost every respect for touch as well. There are elements of tactual texture which are characteristic of a material, and are not affected by the external form of the material. To apprehend the smallest tactual elements, one spontaneously uses the portions of the body most sensitive to touch, the fingertips, or, sometimes, the lips. If one touches a piece of stone or wood, then one feels the typical uniformity of the elements, which yet does not mean complete uniformity. One has a fine feeling for how far the formal elements can vary without destroying the unity of an impression of a

certain material. But, as in vision, here, too, there are structures of tactual elements that are characteristic of the material. The difference between the elements of stone and wood, for example, persist in a characteristic manner in the structures of a higher order. Whether we stick to the smallest elements or to their higher structures while touching, depends upon the general situation as well as the task facing us. As with the visual elements, the tactual formal elements also generally lead a remarkably little noticed existence. They help us the most in recognizing materials, but in a way they completely consume themselves in the process. What was said above about the difference between artificial and natural forms of visual material is also valid, *mutatis mutandis*, for the tactual area. Suppose that a certain pattern, such as a corrugation, is imposed on a wooden surface. One then feels "corrugated wood." The form producing the impression of corrugation is considerably larger in extent than the formal elements producing the impression of wood. However, there are two more differences: the completely regular repetition of the singular element in the case of corrugation, and a certain difference between the finest elements of the wood, in spite of all their general similarity. One is tempted to express the difference with an image from acoustics, by contrasting musical tones and noise. We will see later (Section 45) that more lies behind this image than a mere means of illustration.

When we stick our hand into the loose sand at the beach, and then move our fingers, we encounter a material in a natural form, whose mode of appearance is not identical with any previously mentioned. Neither a surface touch nor an immersed touch in the previously given sense is present. Flour or, generally, any substance consisting of the finest particles, feels similar to sand. Nothing visual corresponds to these tactual phenomena, for although one might believe that he or she is able to see the internal structure of materials like sand and flour, they present themselves phenomenally as surface colors.

Language has not ignored the difference between natural and artificial forms of material discussed here. We tend to speak of thing or object in the case of artificial forms, and of material or substance in the case of natural forms.[2]

Footnotes

1. I consider the perception of surface structure to be just as elemental an activity for the retina as the perception of color or the perception of figure, as in the sense of E. Rubin, *Visuell wahrgenommene Figuren* (Visually perceived figures). Gyldendalske Booghandel, 1921.
2. Editor: Katz apparently has in mind here the distinction between count nouns and mass nouns.

Section 11. Continuity of the Tactual Surface; Tactual Figure and Ground

1. *Visual and tactual continuity.* One question which has intentionally been left aside, but now deserves a brief assessment, concerns the complete filling in of a touched surface with tactual matter. Since a quite comparable issue exists for vision, it will briefly be presented first for that sense. Looking at the paper of this printed page, one would rightly say that it does not contain any visual gap, however small. The entire surface of the paper is covered, completely without interruption, with visual matter. The distribution of the rods and cones on the retina, however dense their placement, allows only discrete stimulation. If the gap between receptors goes unnoticed, then this must be due to a central compensation. The fact that light-insensitive gaps in the retina can be bridged centrally by a completely different dimension is shown by the negative scotoma and, above all, by the phenomena at the blind spot.[1] The latter also reveals the type of compensation, which, as is well known, involves an assimilation to the blind spot of the surrounding sensory contents. I have already called attention (Katz, 1911, p. 294 f.) to the assimilation phenomena of the peripheral retina, both as regards to the assimilation of color matter and to the modes of appearance of color.[2]

The matter of compensation and assimilation can easily be overlooked, if not for the blind spot, then at least for the other portions of the normal retina, because of the minuteness of the spaces between sensory receptors. It is different for the skin senses, where the insensitive gaps are large, even relative to the sensory points to which one attributes sensitivity. In fact, if the five fingers of the hand are considered to be *one touch unit*, as in principle is permissible, then the insensitive gaps between them are simply colossal, and yet nevertheless the gaps are spanned (Section 34). Reversing our transition from eye to skin, Ebbecke remarked: "Rarely do we have a granular temperature sensation or notice the temperature-insensitive gaps of the skin, in spite of the very sparse distribution of the temperature points. Just as rarely do we have a granular color sensation in looking at a colored surface, even though the individual cones in the peripheral retina are spaced rather far apart."[3] In my experience, a completely uninterrupted tactual surface is experienced most clearly with very smooth and hard surfaces, e.g., metal and glass. Here we have only the *experience* of an undoubtedly continuous tactual surface, with no regard given at all as to whether the judgment based upon our experience actually conforms to objective fact, i.e., whether the surface actually is uninterrupted. A general definition of continuity is not called for, since the continuity of a tactual surface can be experienced even where the tactual gaps are undoubtedly clearly visible. There is stiff linen which is so woven that the square openings between the individual threads are nearly two to three times as wide as the .25-mm. thick threads themselves. When touched, such linen usually gives the impression of a rough, but uninterrupted surface, even when it is so situated that the touching fingers cannot feel any underlying surface through the holes. What does this example indicate? That genuine continuity can be experienced where, at least for the eye, gaps exist. One could suppose that the experience of the uninterrupted tactual surface actually derives from the representation of visual continuity, with which it is very closely connected, and that there is no genuine tactual continuity at all. This supposition will become an assured fact when one succeeds in proving that no spatial properties at all adhere initially to the sense of touch (Section 46). This proof cannot be provided by pure introspection, for in our experience, there are, without a doubt, uninterrupted and interrupted tactual surfaces. We first of all have to acknowledge these phenomena as such. Even if everything spatial enters touch only via vision, it still remains to be determined what constellations of tactual stimuli evoke the visual image of continuity, and what evoke that of discontinuity or gaps.

In many cases, where it is important to ensure the continuity of a surface as well as possible, one prefers to rely on the eyes rather than the touch. Why? Because vision has open to it certain ways to assure the impression of continuity that are not available to the sense of touch. One can change the illumination of the surface being examined and its orientation to the eye, and the eye can be moved closer to the surface or even get right up against it with a magnifying glass or microscope. With touch, one can only call upon the most sensitive organs, which nevertheless had usually been entrusted with the investigation in the first place, and for the sense of touch there are no such things as a magnifying glass or microscope.[4]

If I put fine sand in a container so as to produce a completely level surface, then I see a completely uninterrupted surface, which looks similar to sandpaper. When touched, the surface proves not to be continuous, but dissolves into movable grains of sand. In this case, more trust is given to the tactual experience than to the visual impression. This is one of the many cases where the sense of touch takes over the leading role from the eye.

2. *Figure and ground in vision and touch.* In regards to seeing with the blind spot, the predominant view is perhaps that nothing is actually seen with it in the proper sense of the word; it does not assert itself as a gap not covered with visual matter. If we look at a piece of stiff linen, then we do not see nothing at the square openings, but rather empty space that is free of matter. Stiff linen provides an opportunity to study ambiguous figures in the sense of Rubin (op. cit., p. 30 f.). One can either emphasize the square gap as figure, with the thread receding to ground, or, as is probably more natural, allow the space occupied by the thread to serve as the figure, with the openings then becoming the ground. Rubin certainly would not object

if we were to apply his very stimulating reflections concerning visual figure-ground, *mutatis mutandis*, to three-dimensional tactual structures. If you move [your hand, etc.] over the bristles of a stiff brush, without bending them much to the side, you will feel a discontinuous space filled with points, a tactual figure. The points give the impression of a great numerosity, whose estimation might seem a completely hopeless task. Between the points, there is not "nothing" in a tactual sense, but rather empty tactual space that is not covered by matter. The tactual space is covered discontinuously with the tactual matter of the brush points; the space between forms the *tactual ground*. Unlike the case in vision, the tactual figure and ground are not reversible in this or other cases of a similar nature. However, this naturally does not make us give up the distinction between tactual figure and ground. In some cases, the experience of the ground is clearer than with the brush, e.g., in the case of scrapers with iron prongs, such as used in kitchens. In other cases, it is so unclear that one can fail to recognize it at all, e.g., in the case of woven material such as velvet and corduroy. Experiences of ground can always be demonstrated, as far as I can see, wherever the stereognostic ability comes into play in touch, that is, where objects are recognized tactually by this or that property.

While the visual phenomena mentioned here all occur when the eye is motionless, parallel tactual phenomena only become really clear when the touch organ is in motion. However, this difference does not invalidate the comparison made here between vision and touch, nor anything that has been said about the tactual domain.

The investigations cited above studied the connection between the discrete distribution of the visual and tactual receptors and certain spatial properties of visual and tactual experience. We have not hitherto touched at all on how the experience of space *arises*, because that belongs to a quite different level of discourse. It goes without saying that a property of visual and tactual space such as continuity would not thereby be "explained" if the retina and skin were not discretely, but rather continuously sensitive.[5]

Footnotes

1. According to recent studies by K. L. Schaefer, the phenomena at the unstimulated retinal portions in dim light also could be cited; these correspond to those of the blind spot in bright light. *Pfluegers Archiv*, *160*, 1915.
2. Assimilation phenomena in the case of colors have also been studied by W. Fuchs. *Zeitschr. f. Psychol.*, *92*, 1923.
3. U. Ebbecke, Der farbenblinde und schwachsichtige Saum des blinden Flecks (The color-blind and vision-deficient border of the blind spot). *Pfluegers Archiv*, *185*, 1920, p. 178.
4. Editor: In Section 32, however, Katz describes the use of an intervening cloth to aid in the reading of braille; for a review of more recent work indicating the

value of an intervening paper or cloth, see the Introduction above and Krueger's (1982) review. The fingernail also might be used to exaggerate textural cues (S. Lederman, personal communication, June 1988).
5. Statements by R. Pauli in his report on M. von Frey's work on the two-point threshold probably also refer only to the first set of questions touched on by us: "By the continuous flow of excitations into one another is the scattered distribution of the receptors in the skin compensated for to a certain extent and the notion thereby engendered that the sensitivity of the sensory surface is continuously extended." *Arch. f. d. ges. Psychol.*, *48*, 1913, p. 42.

Section 12. Tactual Images of Things: Memory Touches

"For the color in which we have most consistently seen an external object is impressed indelibly on our memory and becomes a fixed property of the memory image. What the layman calls the real color of an object is a color of the object that has become fixed, as it were, in his memory; I should like to call it the *memory color* of the object."[1] Snow, for example, is "in reality" white; it has the memory color of white. If we happen to think of it with its color, then the color we reproduce is white, in spite of the fact that it can transmit many different types of light to our eyes under various conditions of illumination and certainly has done so in the past. What we have just said about snow also holds for the many other objects in our environment that have a characteristic color; they are mentally imaged with the particular color ascribed to them. It now remains for us to investigate whether there is something in touch which corresponds to the memory colors. Are there memory touches? We should not expect a complete correspondence, because, after all, in tactual perception there is no factor corresponding to illumination, but does perhaps at least a partial analog exist?

We begin by asking whether we have tactual images at all of objects we have once touched. Without a doubt, one can call up, more or less clearly, a tactual image, usually accompanied by a visual image, of a surface touched in the past, such as glass, sandpaper, velvet, wool, or wood. I have not yet met anyone incapable of doing this. This finding is no great surprise; why shouldn't the sense of touch be outfitted with images? However, the divergence of the structure of the tactual representation from that of vision deserves our attention for a moment. The bipolarity of tactual phenomena (Section 4) is true in general, and holds in the same way for the tactual image as for the tactual sensation. We imagine a color independently of the sentient sense organ, but a tactual image always bears the trace of the touching agent of the body. If I carefully examine how I imagine something like the smooth touch surface of a pane of glass, then I indeed confirm the fact that I imagine I am moving a touch organ over the glass pane. I can also make myself imagine my hand resting motionless on the glass pane, but then the image of the smoothness is no longer clear. This fact reflects in a noteworthy way the significance (considered in detail below, Section 16) of movement as a formative force in touch. I have never succeeded in producing a tactual representation from which the image of the touching organ was completely excluded.[2]

If you asked many people to render a definite tactual image of glass, emery cloth, silk, leather, paper, linen, wool fabric, wood, fur, or tricot, you would then be able to establish that: 1) without exception, all would imagine with at least one tactual organ, and 2) almost all would think the imaginary touching was carried out with the *fingertips*. I have been told extraordinarily infrequently, even with the questioning made intentionally somewhat suggestive, that the imaginary touching was carried out with a different part of the body, such as the foot, arm, or lips. Now, without doubt, in everyday life we have come into contact with many of the materials mentioned above through other parts of our body than the fingertips, at least for the portions of the body that are constantly clothed, and for that reason it would be expected that our questioning would sometimes elicit the concurrent rendering of a different tactual organ than the hand. By what principle does the memory make its selection, when prompted by the above mentioned task to render a tactual image? One would be inclined to say that memory reproduces the tactual image of a material, together with the tactual organ that has touched the material most frequently, which is precisely the hand with its fingers. This is certainly true for materials like glass, wood, paper, and leather, but not for materials like linen, wool, and tricot, since the part of the body constantly clothed by them is not reproduced along with them. With these latter materials, then, still another principle obviously holds. Of all the tactual representations evoked by having a material touch the various parts of the body, that is reproduced as the *representative tactual image* which has the greatest degree of sharpness, and that is precisely the representation we obtain through the most sensitive tactual organs of the body, the fingers.[3] Here we disregard the very rare touching with the lips, which are as sensitive as the fingers (Section 26). Accordingly, the representation of the touching fingers belongs to the representative tactual image. In addition, since the fingers alone recognize the fineness of the weave in fabrics like linen, wool, and tricot, these materials are imagined with the tactual impression received by the fingers, rather than some other, less effective tactual organ. Very few things are not *also* touched by the fingers, the only exception perhaps being food that is taken directly into the mouth from the plate and first touched there.[4] And thus, it may well be said that the impression received by the fingers provides the representative tactual image for nearly all materials. The kinship between what we call the representative tactual image and Hering's memory color is obvious, and we wish to go further and give it linguistic expression by designating this tactual image as the memory touch. We therefore define memory touch as that tactual image which is mediated by the fingers and which, having the greatest degree of sharpness, is used by memory as the representative for all of the other tactual representations of the material. Our image of the world, therefore, insofar as it is a tactual image, is mediated through the touching hand.[5]

A survey of the devices built to adapt nature to our conditions of existence indicates that the hand reigns supreme. How infinitely varied are the contrivances that surround us, yet all become accessible to our will almost entirely because they provide a place where the hand can grasp them. All our implements, from the most primitive to the most delicate, are made for manual use; almost every machine is built to be operated by the hand; the handles on the doors of houses, automobiles, etc. are—their very name already says it—constructed for the hand. The amputee who has lost one or both hands must reconstruct the domestic environment so as to make its parts accessible to the auxiliary organs—the mouth and the feet.[6] In a world of the armless, everything would doubtless be adapted to the feet instead of to the hands. If we imagine a project requiring muscular effort—let's say we map out a physical or psychological experiment—then we also operate internally with our hand on the imagined objects. Thus, just as the hand is dominant in the world of touch, drawing the touch image into consciousness, so, too, is it dominant in the world of action.

According to Hering, the memory color can influence assimilatively the color impression that we receive from an object, or it can enter into the representation process as a separate, autonomous element. As an example of the first mode of operation, when we think we have grass before us in a particular setting, then it can appear green to us, even if our retina actually is not receiving light rays that customarily produce a sensation of green. Although I believe I have shown that Hering attributed too extensive a role to memory color as an assimilative factor in the perception of color (Katz, 1911, p. 214 f.), one can nevertheless not doubt its effectiveness in many cases. More obvious is the intrusion of the memory color as an independent element in the representation process. We think of snow, and its whiteness appears as an image. Memory color as an independent color image can be elicited in very different ways, probably most readily by the designation of the object itself that bears the color in question. If we turn now to memory touch, it, too, like memory color in the second instance, can enter into the mental image as an independent touch representation. And it, too, can then be elicited, in a manner quite like that of memory color, by the designation of the bearers of the touch in question. Thus, a designation like "sandpaper" evokes in consciousness the representation of the surface touch specific to sandpaper. However, as I have occasionally noted, a memory touch can also be evoked by visual means, as illustrated by the following case of being caught by surprise. On the leaf of a washstand I saw the typical veins of marble, touched them, and was greatly surprised that they showed neither the cold smoothness nor the inelastic hardness that are characteristic of marble. The visual appearance had misled me. I felt that I was dealing with wood which had usurped the elegance of marble by means of a not even very artful coat of paint. My surprise indicates that the specific memory touch of marble had been evoked or at least put into a high state of readiness.[7] Here is another, somewhat different case in point. Seats in a restaurant were covered with a red plush, which gave the sure impression of comfortable stuffing. In reality, the seats consisted only of wood, over which the plush had been drawn without being filled with cushion material. I experienced myself and confirmed in others, not without a certain amusement, that one sits down on the seat expecting to experience a soft volume touch on a certain part of the body, only to be painfully surprised by the hard reception. Here, the visual impression of the plush evoked its memory touch and at the same time the representation of a soft volume touch, such as is attributed to cushions. This case shows, as other observations confirm, that there are corresponding tactual images for volume touch as well.

Can a memory touch also influence a tactual percept in the same way that a memory color influences a color percept, that is, assimilatively? I believe that it can. I watch as I touch my upper arm with a series of touch surfaces, e.g., with wood, paper, stiff linen, leather, and silk. I then believe that I am able to recognize the specific surface touches with full clarity. Since it is easily demonstrated that these parts of the upper arm cannot recognize and discriminate such materials when I shut my eyes and have myself touched by an assistant, it must be that the sight of the touch surfaces evoked the specific surface touches and that this worked a transformation of those nonspecific tactual impressions. I believed that I felt wood, paper, stiff linen, leather, and silk, one after the other, where I otherwise would have had impressions that were quite similar to one another. Perhaps we should also include here the following case, which we will return to below (Section 22) in a different connection. The impression of many specific surface touches occurs only with a moving tactual organ, but not with a motionless one. However, when a motion giving rise to a clearly perceived, specific surface touch is stopped, then one believes that these touches persist for some time longer at the fingers. Perhaps the persistence impels the memory touch to reshape the character of the nonspecific tactual impression of the motionless tactual organ. Another explanation of the "tactual afterimage" involves a positive tactual memory image, for whose occurrence the conditions are particularly favorable. For now, we will not try to decide between these two possibilities.

As already mentioned, even if the tactual image of a surface is spontaneously associated almost exclusively with the fingers, this subjective aspect of the representation can nevertheless still be varied at will. In large-scale experiments, where each participant had to write down his or her observations, I found that, upon a request from the experimenter, a particular tactual image could be attached to nearly any part of the body. However, the different tasks nevertheless seemed to vary in difficulty. One can more readily think of a touch being carried out with the lips or the toes of the foot than with the knee or elbow, more readily with the knee or elbow than with the chest or shoulder blade. Previous use of a part of the body as a tactual organ may facilitate using it again in the imagination as a tactual organ. From the answers I received, one cannot doubt that it is possible to imagine touching materials with parts of the body that they never in all likelihood have been touched with in reality. Who, for example, has ever touched emery cloth with the shoulder blade, and yet, asked to do this in the imagination, most people will succeed in doing so. From what I have established, subjects in this type of experiment believe, regardless of whether or not such a touch likely ever occurred in reality, that they have a quite clearly specific tactual image of a material. This image is more definite than the sensation itself could be at the rather insensitive places in question. This indicates that memory touch finds application in these experiments as well. We must resist the temptation here to pursue additional questions on the tactual imagination, since that would force us too far afield from the main course of our investigation.[8]

The designation tactual image has not and is not always understood in the way used above. We contrast *tactual sensations* (tactual perceptions) with *tactual images* in that the image represents the perception in our mind. Differing from this kind of usage, tactual sensation often has been used to denote the so-called "pure" or simple tactual experience not yet influenced by knowledge and therefore not hinting at the objective nature of the material touched, whereas tactual image has been used to denote the transformed (processed), more complex touch experience already referring to things (objects). For example, statements by E. H. Weber indicate that tactual sensation for him is not a tactual experience that refers to the external world; if such reference intrudes, then we have a tactual image. ("Resistance is not a sensation, but an image"; op. cit., p. 47). Now, tactual sensation and tactual image can certainly be defined just as one pleases, but the definition should not invest these concepts with characteristics that imply a particular theory of perception that requires further assumptions. Who could name a tactual phenomenon that has actually remained completely untouched by experience and that contains no reference whatsoever to the external world? Might one point to something like the sensation that occurs when a pressure spot on the skin is stimulated? Well, then we must call attention to the fact that this sensation is at least localized, something that is not entirely possible without experience. The view that the sensation of a pressure spot is something particularly elementary could only be held by an atomistically-oriented sensory psychologist. We certainly have much reason to suppose that the realm of tactual phenomena, just like that of colors, exhibits a series of layers of various age, whose unraveling is made possible by special reduction processes. It also makes a certain sense to reserve the expression tactual *sensation* for the oldest hypothetical layer thus discernible, but one should not lose sight of the fact that we are dealing here with the formation of a limitary concept, not something that has psychological reality.

Footnotes

1. E. Hering, op. cit., *Zur Lehre vom Lichtsinn* (Outlines of a theory of the light sense), p. 7.
2. H. Henning remarks in his experimental studies on the psychology of thinking, "that tactual experiences in particular are undeniably quite closely connected with our body and the organic sensations. Tactual impressions localized outside of the body never occurred in the present series" (*Zeitschr. f. Psychol.*, 81, 1919, p. 78). As our account presented above shows, that would contradict the essence of the tactual impression.
3. E. Mach once said, in comparing hearing and touch, "that the fingertips correspond to the macula lutea." *Die Analyse der Empfindungen* (Analysis of sensations), 9th edition, 1922, p. 152. Mach was thinking here only of the heightened sensitivity of the sensory areas being compared, but the similarity actually goes further, insofar as the memory touches have their entrée through the fingertips and the memory colors through the macula lutea (Katz, 1911, Section 28).
4. I am reminded here of the Oriental who, during a trip to Europe, asked why Europeans did not put rice into their mouths with their bare fingers, since, after all, the enjoyment begins with touching the soft grains of rice with one's fingers.
5. The mode of imagination of a congenitally-blind person was described in a very sculptural fashion: "In his *right* hand (my emphasis), if he is not left-handed, is the focal point of all his ideas in judging and inferring." K. Buerklen, *Blindenpsychologie* (Psychology of the blind), Leipzig, 1924, p. 182. In the illusions of an arm amputee, the hand is given as the clearest part (Katz, 1921, Section 6).
6. Boehm describes amputee devices which, in part, are really original. *Selbsthilfe der Amputierten, insbesondere der Ohnhaender* (Self assistance in amputees, especially those without hands). Verhandl. d. deutschen orthopaed. Gesellsch., Stuttgart, 1921.
7. On the highly demonstrable power of surprise in psychology, see G. E. Mueller, *Zur Analyse der Gedaechtnistaetigkeit und des Vorstellungsverlaufes* (On the analysis of memory activity and the process of mental imagery), Teil 1 (Part 1), Section 9, Leipzig, 1911.
8. Factually and methodologically, there are certain points of contact between experiments of this type and experiments such as those on visual imagination carried out by G. E. Mueller. *Zur Analyse der Gedaechtnistaetigkeit und des Vorstel-*

lungsverlaufes (On the analysis of memory activity and the process of mental imagery), Teil 2, Abschnitt 5, Kap. 2 (Part 2, Division 5, Chapter 2), Leipzig, 1917.

Addendum. Eidetic Tactual Images and Tactual Hallucinations

When Henning[1] on occasion asks whether all non-abstract images of the lower senses are eidetic, then this must be answered emphatically in the negative, insofar as one customarily includes touch among these senses. Non-abstract are the tactual images spoken about hitherto, but in no way do they have the near-to hallucinatory intensity of eidetic images.

Kroh[2] concluded that tactual perceptual images are much more wide spread than acoustical perceptual images. He carried out a series of experiments on the onset of tactual perceptual images, testing whether perceptual images occurred when the subject was brushed lightly with a dull object two or three times in succession on the back of the hand. The percentage of tactual eidetic subjects (76%) appears so high to me, that I would allow a much greater role for suggestion than Kroh has in the prevalence of positive statements from the young subjects.

Wittmann[3] studied tactual perceptual images in a more systematic and detailed way. As with Kroh, the perceptual images consisted only of simple impressions of contact. As for perceptual images of more complex tactual forms, which favor the objective pole, the situation remains uncertain at present.

My book on amputees (Katz, 1921) dealt in detail with hallucinatory tactual images that are not due to mental disturbances and have been known about for a long time, the so-called phantom limb of the amputee. As a rule, the amputee hallucinates the missing part of the body, or a portion thereof, but very rarely, by contrast, does he or she hallucinate a touched object. The pronounced tendency in this case to favor the subjective pole is also evident in psychopathological tactual hallucinations. With these hallucinations, something usually happens with or on the body of the mentally disturbed person. The hallucination of a *touched object* seems to be rarer than that of a seen object.[4]

Footnotes

1. H. Henning, Assoziationsgesetz und Geruchsgedaechtnis (The law of association and odor memory). *Zeitschr. f. Psychol.*, 89, 1922, p. 39.
2. O. Kroh, *Subjektive Anschauungsbilder bei Jugendlichen* (Subjective perceptual images in children). Goettingen, 1922, p. 145 f.
3. J. Wittmann, Ueber das Gedaechtnis und den Aufbau der Funktionen (Memory and the structure of functions). *Arch. f. d. ges. Psychol.*, 45, 1923.

4. See, for example, G. Stoerring, *Vorlesungen ueber Psychopathologie* (Lectures on Psychopathology), Lecture 4, Leipzig, 1900.

Section 13. Refutation of Possible Objections

It perhaps might be said that the tactual phenomena we have described, unlike those investigated in detail hitherto by science, do not involve elementary perceptual phenomena. The new tactual phenomena are not elementary, but rather smack of something inferential, or, indeed, actually constructed. Since the investigations refer to properties of objects, they no longer pertain to psychology, but perhaps to the theory of objects (*Gegenstandstheorie*).[1] We will examine these objections briefly and may anticipate a certain profit for ourselves from doing so.

"Not elementary" can mean: not found originally in consciousness. Then, for example, whoever would deny an elementary character to the various qualities (*Modifikationen*) and identifying characteristics (*Spezifikationen*) of surface touch would declare that they occur later in the consciousness—of the individual, of the phylum, of the animal—than something like the experience that occurs when single pressure points are stimulated. However, if one wants to make such a comparison at all, precisely the opposite in all likelihood is true. Children, while playing, undoubtedly experience touched surfaces, with their multiplicity of special forms, much earlier than isolated pressure points—unless they happen to have been born to experimental psychologists who subject them to laboratory experiments! And everything indicates that a corresponding situation holds phylogenetically as well.

What about the argument that impressions such as surface touch lack the clear character of genuine sensory perception, but rather, as products of inferential processes, involve something intellectual in contrast to a simple pressure sensation? I ask the reader to move his or her fingers over the page of this book or over his or her desk. Does the surface touch that you experience not have a completely warm-blooded clarity and vividness? Is the experience the product of an inferential process? And now, in order to have an "elementary" experience, touch yourself with a pointed pencil. Doesn't exactly this experience seem more foreign, unusual, problematic, artificial? Thus, for this argument, too, which was supposed to downgrade in a certain respect the tactual phenomena described by us in relation to those treated hitherto by science, closer scrutiny turns it precisely into the opposite.

It will take somewhat longer to deal with the objection that the new tactual phenomena arise from properties of objects in our environment, such as hardness, softness, smoothness, roughness, etc., which, however, fall outside of the field of psychology. We are met here with an objection that must be examined in a fundamental manner, and on its own terms. Actually, wide areas of the psychology of perception bear the character of the theory of objects (*Gegenstandstheorie*) and are therefore, if one wants to so express it, apsychological.[2] Now, would one want to deny psychology such investigations, among which I also include those on the modes of appearance of color? Since the theory of objects (*Gegenstandstheorie*) has not yet been fully detached from psychology, and probably never will be, one certainly will not want to take such a radical position. In addition, it would be erroneous to consider the previously studied structures of the sense of touch to be more psychological and less concerned with the theory of objects (*Gegenstandstheorie*) than those given special attention in the present study. This may easily be illustrated with two arbitrarily-selected examples. If we study the sense of touch in the usual way with the aesthesiometer, then the subject experiences a "pointed object," not simply pressure as a state of sensation of his or her body. One could properly disregard the objective aspect of the experience completely, so long as the only concern, say, was to compare the pressure sensitivity of various points of the skin with one another, using a constant stimulus. Thus, since it was held relatively constant in all the investigations, one could abstract from the objective aspect, but without thereby eliminating it. At bottom, the experience of a "pointed" object makes no less a demand for an analysis than does the structure depicting the nature of a surface touch. Even the two-point threshold, insofar as one wanted to consider it here, also has, in the final analysis, an objective side, "object with two points." An unbiased examination thus permits us to discern an aspect of the theory of objects (*Gegenstandstheorie*) even in the older studies on the sense of touch. Thus, as our last statements indicate, it is no longer necessary on epistemological grounds to sharply distinguish between the older studies of the sense of touch and those presented here.[3]

Footnotes

1. Editor: The Austrian philosopher, Alexius Meinong, a student of Franz Brentano, was known for his theory of objects; his essay, *Ueber Gegenstandstheorie*, which first appeared in 1904, was translated as "The theory of objects" in R.M. Chisholm (Ed.) (1960). *Realism and the background of phenomenology* (pp. 76-117). Glencoe, IL: Free Press. He opposed traditional metaphysics' "prejudice in favor of the actual" by claiming that nonexistent objects (e.g., round squares, golden mountains) are objects nonetheless and that each has a definite character (*Sosein*), which is independent of its being (*Sein*). A round square, for instance, has a contradictory character that precludes its being. Meinong's theory influenced phenomenology through the thought of Edmund Husserl.
2. "More and more does one come to the conviction that apsychology is of the same significance for psychology as psychology is for the branches of

philosophy connected with psychology." V. Benussi, *Psychologie der Zeitauffassung* (Psychology of time comprehension), Heidelberg, 1913, p. 497.
3. Thus, there is a transcendency not only for thought, as T. Erismann has recently pointed out emphatically, but for perception as well. *Die Eigenart des Geistigen* (The character of the intellect), Leipzig, 1924.

Editor's Notes on Chapter III, Division I: Movement as a Formative Factor in Tactual Phenomena (Sections 14 to 17)

In the next chapter, Katz develops his views on the importance of movement of the touch organ relative to the stimulus. He first criticizes previous investigators for favoring temporal atomism (Section 14) and using stationary stimuli (Section 15), which he concludes are not adequate stimuli for the sense of touch. On a more positive note, he explains the need for movement in judgments of roughness, hardness, and elasticity (Section 16), and covers the compensation for movement which evidently is needed to preserve the shape of objects felt piecemeal by the hand (Section 17). In moving the hand and arm over an object, not only does the impression persist as the object moves to less sensitive areas of the arm, but the object maintains its identity qua object, as successive sets of individual receptors are triggered, and it is felt to maintain a fixed position in space as the arm glides by it.

Except for elasticity, the final percept is devoid of any trace of movement; the sensory flux is discarded once the important information has been extracted (Gibson, 1966, 1979). Thus, attention is firmly fixed on the invariant properties perceived by means of the motion, not the motion itself. It does not matter whether the motion is in the hand or in the stimulus. What is important is not the motion per se, but the emergent properties or relationships (roughness, hardness, shape, etc.) revealed by the series of successive stimuli. A moving stimulus simply provides more information. A more sensitive sense organ would obviate the need for movement, and, indeed, in the

Addendum to Section 17, which is devoted almost exclusively to the eye, Katz proposed that the general trend in development is from the cinematic to the static. Motion perception in human vision is largely relegated to the peripheral retina, he noted, thus allowing the fovea to do a superb job in apprehending motionless stimuli, a feat unknown in many lower animals, which respond mainly to moving stimuli.

Chapter III:
Movement as a Formative Factor in Tactual Phenomena

Section 14. The Bias Toward Temporal Atomism in the Approach of Previous Studies on the Psychology of Perception

Surely, almost all of the tactual phenomena treated in Chapter II differ from the phenomena in other sense domains, in that to produce them the sensory organ and the stimulus must move with respect to each other. Perception of immersed touch presumes a movement of the hand or the medium just as does perception of volume touch. Surface touch, with all its varieties, also can be recognized and distinguished only through movement. The importance of movement in at least perfecting, if not actually making possible, so-called stereognostic performance and many other activities of the tactual sense, which are now treated here, definitely did not escape the notice of earlier researchers. However, a special attitude toward sensory perception, which we will term "temporal atomism," prevented them from recognizing the full significance of movement as a formative force for tactual phenomena. E. H. Weber devotes a special section to the perception of the shape and distance of objects through intentional motion of the limbs. Among other things, he states: "Overall, it is scarcely to be believed how much we depend upon the intentional movement of our limbs in perceiving the shape of objects, their surface texture (roughness or smoothness), their hardness and softness, and their distance from each other. Close your eyes and rest your hand on a good support. If various pieces of glass, metal, paper, leather, or some other substance are now brought into contact with and moved past the fingertips, then you will confuse substances with each other that you would immediately distinguish if you moved your hand (op. cit., p. 90)." Experiments presented below (Section 23) will show that in moving the touched surfaces with respect to the resting hand, the performance need not decrease so markedly as Weber depicts it here, but we will disregard that for now and consider only his correct report that movement is astonishingly important for tactual performance. Helmholtz (1879) had the stereognostic quality more in mind in his following remarks: "What we are able to determine through the skin sense when we quietly lay our hand on something, such as the imprint of a medal, is extraordinarily dull and lacking in comparison to what we find out by a touching movement, even if only with the tip of a pencil."[1] Like Weber, Titchener in the quotation cited above (Section 3) connected the tactual phenomena of smooth and rough with the movement of the touching organ, and in so doing he distinguished between continuous and intermittent movement. He also considered movement to be necessary for the perception of oiliness. In the quotation from Thunberg (Section 3), the need for movement in the perception of surface texture is pointed out in the same way as in Weber's case, using almost the same examples. In light of such statements, one might well wonder why no one has pressed on to emphasize movement as fundamentally an *elementary formative factor* in tactual phenomena, as well-nigh as indispensable for touch as light is for color sensations.

This failure can be shown to be closely associated with the research orientation or attitude that we characterized above as atomistic. The atomism of sensory psychology in its histological approach, with the goal of determining the function of sensory elements, is paralleled by an atomism in the temporal domain. The two biases reinforce each other in their effect of excluding from investigation many relatively complex cases of perception. The temporal atomism is not limited to the sense of touch, but can be pointed out in every sensory domain. Thus, a "tachistoscopic" bent runs through the entire methodology of physiological optics. The tachistoscopic method is employed like a kind of temporal magnifying glass in order to separate the components of the percept from their groupings, or at least to reveal the points of their connection in the overall structure and to make visible the structural elements of sensation. The briefer the effective duration of the stimulus, the stronger the effect. When one works with longer stimulus durations, the temporal atomism assumes a different form, that of a motionless stimulus. As a rule, the stimulus is kept as stationary as possible during the entire period of stimulation on the assumption that movement of the stimulus creates nothing new, apart from the impression of movement itself, and that the atoms of sensation simply summate over time. Certainly the tachistoscopic method has its justification and the integration procedure mentioned above is permissible in almost all areas of psychological optics as well as in other sensory domains. As far as the sense of touch is concerned, however, we intend to show that processes occurring over time result in conscious phenomena that by no means can be conceived of as the sum of those experiences which arise in connection with motionless tactual stimuli.

It is indeed not coincidental that we are able to cite here three strongly philosophically-oriented psychologists who have spoken out on occasion against the indicated temporal atomism. Cornelius called attention to how the impression of roughness obtained by touching a surface ceases as soon as one notices the individual, successively-touched bumps as such. "Noticing the multiplicity of sensation in succession is no self-evident process; only rarely, and by special concentration of the attention, are the successive points within a short period of time perceived individually."[2] We will find Meinong's statement,[3] "that there are mental images whose characteristic quality requires a period of time in order to be developed," to be completely confirmed in the area of touch. Finally, W. Stern spoke out emphatically against temporal atomism: "The notion that the totality of consciousness can encompass only such contents as are simultaneous and present together at some time ... is a dogma which, with more or less elaboration, dominates much of psychological thought. I consider the dogma ... to be false."[4]

In memory research, where one deals with successive complexes as phenomenal units, a break is made with the temporal atomism attitude.[5]

Footnotes

1. H. Helmhotz, *Die Tatsachen in der Wahrnehmung* (Facts in perception), Berlin, 1879, p. 19.
2. H. Cornelius, *Vierteljahrsschr. f. wiss. Philos.*, 12.
3. A. Meinong, *Zeitschr. f. Psychol.*, 6, 1894, p. 448.
4. W. Stern, Psychische Praesenzzeit (Moment of consciousness), *Zeitschr. f. Psychol.*, 13, 1897, p. 326 f.
5. G. E. Mueller, *Komplextheorie und Gestalttheorie* (Complex theory and Gestalt theory) Goettingen, 1923, p. 4 f. W. Peters has recently spoken out in general against atomism in the psychology of memory. *Zeitschr. f. paedag. Psychol.*, 25, 1924.

Section 15. The Stationarity Principle in the Methodology of Earlier Tactual Experiments

If one examines the methodology of earlier studies on the skin sense, mainly those of M. von Frey and his school, which were unsurpassed in deftness of method, one finds unintended, but unmistakable, indirect evidence that only the moving stimulus and not the motionless stimulus is an adequate stimulus for the sense of touch. This evidence will be presented here. Von Frey and Metzner showed that with successive stimulation, two adjacent pressure spots, each of which is excitable by an isolated stimulus, can be distinguished so that the successive threshold can be appreciably equated with the mean distance of these receptors. "It does not prove, as Judd states, that Weber's method is useless or less reliable, but that a different threshold is determined by each of the two procedures."[1] The tremendous superiority of the moving over the motionless stimulus for the two-point threshold, as we may certainly characterize the result thus obtained, bespeaks of the superior biological significance of the successive threshold. Rupp stated the matter quite correctly (op. cit., p. 158): "Unfortunately, investigation of the tactual threshold has almost always been carried out with the simultaneous imposition of two points. The successive threshold is of at least equal value; it provides finer thresholds and shows only the outermost limits of our ability to distinguish." The successive threshold provides the sole authoritative measure of the practical capability of the sense of touch. To study the sense of touch at rest is almost like wanting to determine the capability of the leg musculature after the leg has been placed in a plaster cast.

How very antithetical to the nature of the sense of touch is a complete lack of motion in the touch organ, can be seen from the fact that von Frey and his students believed that they could only eliminate the smallest involuntary movements of the touch organ, which for them represented sources of error, by placing it into a hollow plaster cast that prevented any movement. It is well-known that in many blind subjects, the natural tendency to move the touch organ (tactual twitches), is so heightened that the organ is difficult to use for the exact determination of simultaneous two-point thresholds.[2]

Detailed investigations by von Frey and Goldmann[3] have demonstrated that the sensation of simple contact with a motionless stimulus fades very rapidly and becomes unnoticeable. This presumably is the result of rapid adaptation. Another consequence of the rapid adaptation of the pressure sense is the oft-cited fact that the clothes covering our body are remarkably unnoticed tactually as long as they do not move. One must be careful not to misinterpret such observations: the touch organs fail to work after a short period in the presence of stimuli which are motionless or which come to rest. On the other hand, they adapt to a remarkably limited degree in the presence of moving tactual surfaces. For example, a cloth band on a rotating disk can be moved over a fingertip with moderate speed and pressure for minutes on end—indeed, for hours if one has the requisite patience—without one being able to establish with certainty any essential change in the touch impression. What else does this mean than that in this investigation the tactual spots adapted in a way that could barely be measured, so little that one reflexively thinks of the "inability to adapt" of hearing? To be sure, one can ultimately deaden the touch organs even with *moving* tactual stimuli, when very vigorous grasping is used (Section 28, Subsection 1), but except for that sort of artificial case from the laboratory, the sense of touch can be regarded as virtually indefatigable for moving stimuli. Can one think of any more convincing proof for the thesis herein asserted, than the fact that the biologically adequate form of stimulus for the sense of touch is the *moving* stimulus and that all experiments using maximally *motionless* stimuli, as productive and valuable as they have indubitably been, have nevertheless only been able to encompass abiomorphic tactual structures (Section 2)? The foregoing excerpts from earlier investigations may suffice to affirm the theory whose proof we stated at the beginning of this section we would provide.

Footnotes

1. M. von Frey and R. Metzner, Die Raumschwelle der Haut bei Sukzessivreizung (Two-point threshold of the skin with successive stimulation), *Zeitschr. f. Psychol.*, 29, 1902, p. 161.
2. On the question of tactual twitches, see e.g., S. Heller, *Entwicklungsphaenomene im Seelenleben der Blinden und ihre Konsequenzen fuer die Blindenbildung* (Developmental phenomena in the mental life of the blind, and their consequences for education of the blind). Berichte der Blindenlehrerkongresse (Report from the Congress of Teachers of the Blind), 1904.
3. M. von Frey and A. Goldmann, *Zeitschr. f. Biol.*, 65, 1914.

Section 16. Movement as a Creative Force in the Sense of Touch

If the preceding discussion has demonstrated that motion emerges as an important means for intensifying the action of motionless stimuli in the touch domain, without changing essentially the nature of the phenomena thereby engendered, then other analyses indicate that motion can create touch phenomena that are not at all accessible with motionless stimuli. As we will see, the sense of touch characteristically reacts to *successive* stimulation with structures to which nothing of motion in any form appears to belong. It is as if the cinematic form of the stimulus is converted to the static properties of an object.

The fact that motion creates touch phenomena that exist only by the grace of motion, is demonstrated by all qualities (*Modifikationen*) of surface touch. Smoothness and roughness occur not at rest, but really only when the touch organ moves with respect to the touched surface. The prototype for qualities such as visual smoothness and roughness, may be provided by the sense of touch. The congenitally blind, in any case, receive their knowledge of those impressions only through the sense of touch. If movement is prohibited, then we lose the entire plentitude of qualities (*Modifikationen*) and identifying characteristics (*Spezifikationen*) of surface touch that are reported in this work. However, all of the other tactual phenomena described in Sections 6 to 8 are also coaxed into existence only by movement of the touch organs. Although objective movement is essential for producing these tactual forms, subjective movement enters very little as a component of their appearances. In order to completely grasp the special qualities of what I am trying to describe here, I must again ask the reader to meet me halfway by producing a few surface touches himself or herself. Make a movement over your deskpad, and feel its quality of "soft roughness." This impression builds up during the movement, without having even the slightest trace of movement as a component in itself. Soft roughness characterizes the deskpad in the same sense as does its color, and one can as little assign movement as a component to the color impression as to the tactual quality. To be sure, one subjectively experiences the movement of the touch organ when it seizes upon the texture of a surface, but the movement does not thereby become a component of the surface touch. One can let his or her eyes wander over objects and apprehend their successive colors, but does movement thereby become a component of the colors of the objects? We will not pursue the phenomenological analysis further at this point; it requires a more specific investigation, which will follow below (Division III).

In addition, all impressions on the *hard-soft* dimension owe their existence exclusively to *successive* stimulation. Resting the touch organ really motionless on an object precludes any sure judgment as to whether it is hard or soft. The fingertip has the greatest sensitivity for these impressions, whose degree depends upon the extent and temporal organization of the stimulation on the pressure receptors. It is astonishing how varied are the experiences that emerge from variation of the spatial-temporal pattern of stimulation of the finger in its flattening out.[1] In active touch, a greater sensitivity develops if at a certain level of exerted pressure the sensations of the muscle sense [or kinesthesis] also enter in and participate in the development of the impressions. But since judgments of hardness and softness are possible even when the objects are moved against a *motionless*, rigidly-supported touch organ, receptors in the skin must therefore be the actual bearers of the impressions discussed here. Hardness and softness can best be judged at a certain speed of touching movements, but if extremely high or low speeds are excluded, the impression of hardness or softness is invariant for intermediate speeds. Thus, just as the experiences of smoothness and roughness appear to deny their origin in *successive* stimulation, since no temporal components adhere to them as essential features, so nothing more of a temporal nature appears in the experiences along the hard-soft dimension either. If the touch organ comes to rest on an object, then the previously developed impression of hardness or softness persists very briefly (Section 12 gives an analogous case), but this impression, with its somewhat lively quality, then gives way to that of a dead pressure experience.

Finally, the dependency of the impression of elasticity on *successive* stimulation must now be considered. If we press on a spring or on a stretched rubber band with our finger, then we apprehend clearly the elasticity of these objects. We do not merely infer them in an abstract manner. Whereas physics has a unitary concept of elasticity, the phenomenal guises that elasticity assumes are very numerous. We readily distinguish the elasticity of a rubber band from that of a steel spring. We also distinguish an extraordinarily large number of levels of intensity in each form of elasticity. By adjusting the tension of a rubber membrane or the stress on a spring, the experience of elasticity can be set to any desired level. Once again, the spatial-temporal pattern of excitation of the pressure and muscle sense organs is crucial for the development of the experience of elasticity. However, the muscle sense appears to participate much more in this case than in obtaining impressions of hardness and softness. Two things indicate this, first, the fact that one cannot quite get an impression of elasticity by moving an elastic object against a *motionless*, rigidly-supported organ, and second, the fact that one can get very pronounced experiences of elasticity even when the pressure sense is largely excluded. Put a large piece of rubber between your teeth and bite on it. The experience of elasticity can then be very pronounced, yet there is no excitation of the pressure sense to speak of here, at least when little force is used. Though we are accustomed to carrying out the movements to determine elasticity at very definite speeds, we are not rigidly committed to those speeds, just as is the case for the movements to determine hardness and softness. Even if there are certain transitions between the impressions of elasticity and the hard-soft dimension, the two dimensions ought nevertheless as a rule to be sharply distinguished. In our experience, nearly all points on the hard-soft dimension can occur in conjunction with all forms and levels of elasticity. Here, too, our astonishing psychological power to shape the peripheral sensory data is evident.

The literature contains few studies on the experience of elasticity. Some pertinent observations are found in Lotze. "The unyielding stone below our feet causes a different feeling from the wooden step of a staircase or the rung of a ladder.... By the distinctions of the vibrations we can easily tell whether the round of the ladder is broad or narrow (op. cit., p. 589)." In investigating the question of the so-called hardness of animal tissue, Gildemeister conducted some very noteworthy experiments with stretched cords and elastic strings, in which great sensitivity to differences was found.[2] He remarked: "It is actually a ... psychological problem as to what we usually understand by hardness." Two Americans have conducted experiments concerning the nonvisual perception of the length of rods whipped back and forth by the observer, finding that what above all is decisive in judging between two alternative pressure sensations is the change in frequency and intensity of swing produced by varying the length of the weight and the center of gravity of the rods.[3]

People sometimes say that they can see elasticity (or brittleness) in an object. To be sure, one can say something like this, but it must not be overlooked that the experience of elasticity is rooted in the tactual or kinesthetic domains, which provide the interpretation of the visual impressions. Very compelling is the invitation to empathize with the elasticity that trees which move in the wind, such as the birches, present to our eyes. As the fisherman's rod is flexed back and forth, we observe the breadth of its movement, as well as the speed of the movement of its individual parts, and thereby develop a notion of its elasticity, that is, we reproduce the corresponding dimension of tactual-kinesthetic experiences. The tactual-kinesthetic primacy in the formation of elasticity is indicated by the fact that visually-perceived elasticity is subject to verification by our sense of touch, but not vice versa. Even the congenitally blind possess the experience of elasticity and the corresponding representations. Furthermore, our representation of elasticity itself points to *successive* stimulation as triggering the impression of elasticity; one cannot imagine something elastic without thinking about carrying out a movement on an elastic object.

Footnotes

1. M. von Frey thinks there is no deep pressure sensation. "It is . . . recommended . . . that the so-called deep pressure sensations be termed what they are, transformed pressure sensations, stretch sensations, or deep dull pain." *Zeitschr. f. Biol.*, *66*, p. 432. A. Goldscheider has repeatedly (and recently together with P. Hoefer) applied some very noteworthy arguments against this viewpoint. Apart from the actual pressure sense with its pressure points, according to these two authors, the skin and the deeper tissues have a sensitivity for mechanical stimulation that is intended primarily for the perception of changes in the state of the tissue itself. It matters little for our investigation how this issue is resolved, because the disputed deep pressure sensation is not especially significant for the world of touch. Even Goldscheider and Hoefer themselves take the view that the normal pressure sense, and not the deep pressure sense, serves primarily in the recognition of the external world. The deep pressure sensation "provides pressure-like sensations of a dull nature, which cannot be intensified to form strong and hard pressure sensations." Ueber den Drucksinn (On the pressure sense), *Pfluegers Archiv*, *199*, 1923, p. 619.
2. Gildemeister, Ueber die sog. Haerte tierischer Gewebe und ihre Messung (The so-called hardness of animal tissue and its measurement), *Zeitschr. f. Biol.*, *63*, 1914, p. 187.

3. Erna Shults, On the non-visual perception of the length of vertically whipped rods, pp. 135-139, and A. S. Baker, On the non-visual perception of the length of horizontally whipped rods, pp. 139-144; *American Journal of Psychology, 33,* 1922.

Addendum. Movement as a Creative Force in Other Sensory Domains

As far as I can see, a transformation of a set of successive stimuli into a phenomenon that bears no further trace of movement, and which therefore is comparable in certain respects to the phenomenon just treated, only occurs in the case where a particular (motionless) localization of an auditory impression in space results from successive acoustical stimulation of the two ears.[1] At first glance, it appears that much of W. Stern's work mentioned above on the psychological present should be mentioned here. In reality, Stern wanted only to demonstrate that the present consciousness cannot be conceived of as temporally punctual. The problem which concerns us here (how time can be consumed, as it were, in the production of new types of perceptual representations) was far from his thoughts.

What about the myriad facts from the perception of movement? They do not belong here, but they all nevertheless point to the subjective articulation of time as an aspect which is important to them. To be sure, there are phenomena here where the temporal structure as such recedes markedly and movement produces a new type of effect, namely, a high level of spatial enhancement. In monocular stereoscopy,[2] a sufficiently rapid succession of drawings showing an object from different viewpoints, arouses the highly plastic impression of a (moving) body. Similar to this are the cinematographic presentations which have been filmed in movement. In addition, a passing glimpse increases the clarity of empty space, as demonstrated in the well-known observations by von Jaensch on the psychology of space.[3]

I conclude these remarks with an allusion to some little noticed processes in the enjoyment of sculpture, whose explanation probably lies in the same direction. In a discussion, following a lecture by H. Cornelius on the artistic perspective in architecture and sculpture, O. Wulff said: "The artistic creation wants to offer the whole to *simultaneous contemplation* . . .Many classical and modern works of art deliver up their full plastic content not in the main view, but only while *walking around them* (emphasis added), e.g., the Borghese fencer, the Pasquino and other groups.... One cannot even speak of several separate views in these cases, but rather one view glides into another. This is especially the case with many types of early Gothic works, which definitely need to be caught in the movement of walking past.... It follows from this that in sculpture (as in architecture, by walking through spaces . . .) the absorption of the work of art by the observer sometimes occurs not at one time, but over a successive lapse of time."[4] From my own experience, I can only concur with Wulff's statements.

Footnotes

1. According to the theory by E. von Hornbostel and M. Wertheimer, *Ueber die Wahrnehmung der Schallrichtung* (Perception of the direction of sound), Sitz.-Ber. d. Akad. d. Wiss., Berlin, 1920.— According to von Hornbostel, in localizing sound, "the entire, temporally extended oscillation is to be regarded as the stimulus in each case, not merely the state of oscillation at one moment of time." Physiologische Akustik (Physiological acoustics), *Jahresber. ueber d. ges. Physiol.*, 1920.
2. See M. Straub, *Zeitschr. f. Psychol.*, *36*, 1904.
3. E. R. Jaensch, Ueber die Wahrnehmung des Raumes (Perception of space), Erg.-Bd. 6 (Supplement Volume 6), *Zeitschr. f. Psychol.*, Leipzig, 1911, Chapter 6.
4. *Bericht ueber den 1. Kongress fuer Aesthetik u. Kunstwissenschaft* (Report on the First Congress for Aesthetics and Art), Stuttgart, 1914, p. 268. Recently, people have begun to film early Italian sculptures, African sculptures, Buddhist statuary, etc., which are placed on a pedestal that turns on its own axis. When shown on the screen, these works of art are said to produce surprisingly impressive effects.

Section 17. Constancy of the Position and Properties of Objects with a Moving Touch Organ

When we move our eyes, the array of objects glides over the retina. This gliding, however, does not make the objects themselves appear to move. Rather, for us they keep their same positions in space. If the eyes are moved involuntarily, however, then the gliding of the retinal array is attributed to movement of the objects. Apparent movement likewise occurs if the observer believes she has moved her eyes, but in reality the eyes have remained motionless (eye muscle paralysis). Hering and Mach have expounded the theory that movement of objects is not seen if the objective movement of the retinal image corresponds in magnitude and direction to a centrally-triggered counter-impulse. Even the apparent motion seen in the case of eye muscle paralysis can be explained in terms of that theory. Mach (op. cit., p. 111 f.) has already suggested that one can find quite similar situations in the tactual domain as in the visual domain; we now will examine these situations. If I move my hand over a fixed object, e.g., the corner of my chair, then different and constantly changing parts of the hand come into contact with the corner, so that the tactual array glides over the touching surface just as the visual array glides over the retina. Now for us the corner persists just as unaltered in its position in tactual space as does a visual object in its position in visual space when the eye is moved. Therefore, just as with the motionless visual scene perceived in the presence of a gliding retinal array, here, too, we must assume a compensation, due to central movement impulses, which results in a motionless tactual field perceived in the presence of a gliding tactual array.[1] The negative after-effects of movement, often studied in vision, exist in touch as well, according to Thalman.[2] The case of one touch organ touching another touch organ has no parallel in the visual domain. The moving touch organ feels the unmoved organ as the object, as Weber (op. cit., p. 111) has already also pointed out. Only with some effort is it possible to experience with the unmoved touch organ the gliding past of the moving organ. This apprehension no doubt comes from practical experience; movement of a touched surface relative to a touch organ is produced, as a rule, by movement of the touch organ.

No less remarkable than position constancy is the preservation of the shape as well as color and texture of objects during movement of the eyes. Whereas others, e.g., Mach, have examined in detail the preservation of shape, less has been done to make clear the counter-intuitiveness of the preservation of color. The color of a casually observed object hardly seems to change as color-insensitive portions of the retina replace more color-sensitive portions during an eye movement. And completely unnoticed has been the preservation of texture. The texture of objects seems to remain fundamentally the same when moved to regions of very different sensitivity (Katz, 1911, Section 28). The sense of touch shows very similar effects. If we move our hand over an object, e.g., a key, then the form remains constant for us, even though parts of the hand with very different sensitivity are used one after the other in touch. Even more striking is the fact that the micromorphic [substance] properties [e.g., roughness, hardness] of the object also do not change noticeably when the hand moves over it, in spite of the changing sensitivity of the parts of the hand.

When the eye scours the visual field with movements, nothing is recognized during the movements themselves; they are of significance only in bringing the different portions of the visual field one by one into the position of clearest vision. Actual recognition occurs only in the motionless interval between movements.[3] In this respect, the anesthesia of the moving eyes is sometimes spoken of, and some authors have taken the radical position that we are blind during the movement of the eye. However, that is not the case, but rather, as I have demonstrated elsewhere, there is an unarticulated color impression of considerable intensity during the movement of the eye; only nothing is *recognized* (Katz, 1921, p. 27). Visual objects, with their colors and their macromorphic and micromorphic properties, are accessible only to the motionless eye. The reverse holds for touch. It was the task of this chapter to demonstrate in a general manner that the full richness of the palpable world is opened up to the touch organ only through movement. Only through movement does touch get objects to reveal their qualities, whereas the eye throws things into a sensory chaos with its movement. In contrast to vision, touch suffers a partial anesthesia when it goes over to a state of stillness.

Footnotes

1. The movement which results in a shift of the tactual array is sensorially much clearer than the movement of the eyes.
2. A. Wellington Thalman, The after-effect of movement in the sense of touch, *American Journal of Psychology, 33*, 1922, pp. 268-276. The stimulation is produced by having endless bands move over the skin. The apparent reverse movement or negative after-effect is clearer if the band remains in contact with the skin after the movement has ceased. The illusion increases as the band becomes broader, rougher, and is impressed for a longer duration.
3. B. Erdmann and R. Dodge, *Psychologische Untersuchungen ueber das Lesen* (Psychological study of reading), Halle, 1898, Chapter 1.

Addendum. Kinetic and Motionless Figures for Humans and Animals

The importance of movement for the efficiency of touch raises the question as to whether moving stimuli once had a greater significance for vision as well. It is well known that the fovea far surpasses the peripheral retina, except in the ability to perceive movement. This suggests that the peripheral retina, with its bent towards moving stimuli, reproduces a capability previously provided by the fovea. Consistent with this hypothesis is the fact that, according to Stern, "the differential sensitivity of the fovea is greater for moderately fast moving objects than for motionless objects, other things being equal."[1] Observations in child psychology do not provide firm support for our hypothesis; the fact that younger children are stimulated more markedly by moving stimuli than by motionless ones is susceptible to alternative interpretations.[2] However, innumerable observations of animals speak in favor of the hypothesis. Throughout almost all of animal psychology, one finds evidence for the great biological significance of visually-perceived movement. According to Exner,[3] the insect eye is ill suited to apprehend spatial structure, but is well suited to interpret movement. The beetles of the Cerambyx genus, which play dead when touched, presumably have more chance to escape the attack of their enemies when they are motionless. Frogs, lizards, and salamanders[4] take in only moving, not motionless food. We need not consider at all how *conscious* the animals are of the moving stimuli; what matters is that in this crucial situation their visual system responds quite differently to the total stimulus extending over a relatively long time span than to its fragments.[5] This statement holds not only for insects, reptiles, and amphibians, but also has been confirmed in one form or another for all mammals in which vision has not atrophied. In his descriptions based on the most intimate observation of animals, Seton Thompson states time and again that when two hostile animals meet, that animal is victorious which is able to stiffen to a statue sooner, before the other can notice it. Incidentally, in the numerous cases of inborn or acquired adjustment of visually-guided animals to one another, as shown in the recognition of a "definite member of the species," or as an "individual of the opposite sex," or as "prey," or as a "deadly enemy," etc., different kinetic configurations specific to animals are perceived, not simply movements per se.[6]

It is a far cry from the perception of a kinetic configuration by an animal to the perfection shown by the human in the recognition and analysis of motionless visual figures. The fovea of the human, by its capability in comprehending kinetic configurations as well as motionless figures, shows traces of a hastily completed stage of development, at which many animals are kept for a prolonged duration. If I concur with the viewpoint of those, such as Exner, Koffka, Koehler, and Wertheimer, who see something completely new and almost sensation-like in the perception of movement, compared with the impressions of objects at rest,[7] it is because I give considerable weight to the evidence from behavioristic studies in animal psychology as well as that from introspection. I rather believe that an extension of the special hypothesis put forth here to other sensory domains would reveal a general trend in development from the cinematic to the static in sensory perception.

Footnotes

1. L. W. Stern, Die Wahrnehmung von Bewegungen vermittels des Auges (Perception of movement by means of the eye), *Zeitschr. f. Psychol.*, 7, 1894. According to Helmholtz (*Physiologische Optik* [Physiological optics], 2d ed.), a brightness difference of 1/131 can be discriminated for moving stimuli, compared with 1/100 for motionless stimuli.
2. K. Koffka asserted "that the newborn has less ability to see movement than the adult." *Die Grundlagen der psychischen Entwicklung* (Basis of mental development), Osterwieck, 1921, p. 45. Insofar as the ability to see and resolve movements correctly is involved, this may be correct. But insofar as the excitability of visual attention in younger children is concerned, an event that occurs moderately rapidly has a much greater effect than a motionless object.
3. S. Exner, *Abhandl. d. Wiener Akad. d. Wiss.* (Proceedings of the Vienna Academy of Sciences), Section III, Vol. 72, 1875, p. 165 f.

4. "If we stop the movement after the salamander has taken its first step forward, then instead of proceeding further toward the object, which lies openly available to its eye, the salamander stops and slips into rest if we do not provide any new movement. . . . As soon as the worm comes to rest, it becomes at the same instant precisely as indifferent for the eye as if it had never shown movement. E. Matthes, Die Rolle des Gesichts-, Geruchs- und Erschuetterungssinnes fuer den Nahrungserwerb von Triton (The role of vision, smell, and vibration sense in food acquisition in triton), *Biol. Zentralbl., 44*, 1924, p. 77.
5. We must hereupon assume general physiological processes that extend over certain periods of time. However, along with V. Benussi, *Arch. f. d. ges. Psychol., 36*, 1916, p. 64, I refuse to see in them an *explanation* for general mental processes.
6. Everyone knows about the individual kinetic configurations of people. "Nothing characterizes a man better than the manner in which he moves. . . .The expression of a face remains vivid in consciousness even when its image has already faded." E.M. von Hornbostel, Musikalischer Exotismus (Musical exoticism), *Musikzeitschrift Melos,* Jahrg. 1921, No. 9.
7. That is also supported by V. Benussi for the sense of touch.

Editor's Notes on Chapter I, Division II: Studies on Surface Touch (Sections 18 to 30)

In the next chapter, Katz begins presenting his own experimental data, focusing on properties (mainly roughness) of surface touch. In many ways, his experimental methodology does not meet current standards. He typically did not manipulate the physical properties of his stimuli, but rather used what materials were commercially available, and consequently he does not present precise threshold values. When he measured recognition time (Section 29), he included incorrect trials with correct ones, rather than averaging the correct ones separately. The results are also dubious because of the great variability in the recognition times for various materials. He used only a few subjects in each experiment, and did not use "same" pairs as catch trials in the discrimination task. Notwithstanding these limitations, Katz presents richly suggestive findings obtained with a host of very imaginative procedures and apparati.

Part I (Sections 18 to 28) deals with the qualities (*Modifikationen*) of surface touch, such as the rough-smooth and hard-soft dimensions. Part II (Sections 29 and 30) deals with the identifying characteristics (*Spezikationen*), which specify the material (paper, leather, metal, etc.). The latter are the first properties to be lost when conditions become more difficult (reduction of first degree; Section 21); only when conditions worsen considerably are the qualities of surface touch lost as well (reduction of second degree). Part I, which deals with the (more general) qualities, uses a series of 14 papers (Section 18). Part II, which deals with the (more specific) identifying characteristics, uses a wider range of 27 materials, which are not restricted to papers alone (Section 29). Katz found that materials such as kid leather, cloth, rubber, and paper are readily confused when felt by a blindfolded person, indicating that vision, not touch, makes them "feel" different in everyday life. Katz also found that recognition through touch could be considerably aided by allowing subjects a fleeting glimpse beforehand of the objects presented (Section 29).

Reduction procedures are sometimes employed in visual studies. For example, film or aperture color is obtained instead of surface or object color when a homogeneous stimulus is viewed through a reduction screen. (A reduction screen is a screen with a small hole in it.) Katz used many reduction procedures in the present tactual studies, in part perhaps because performance often was nearly perfect otherwise. He used both objective (object-oriented) and subjective (person-oriented) reduction procedures. The latter involve procedures such as preventing lateral movements (Section 22) and adapting the touch organ (Section 28), and thus do not involve "subjective" in the more usual, phenomenal sense. Reduction also was achieved by reducing the size of the stimulus (Section 20), using a veiling intermediary such as adhesive tape or dried collodion (Section 24), and permitting only a very light pressure (.1 g) on the touch surface (Section 29).

Katz also varied the difficulty of the task by testing body parts not normally used to touch with, such as the lips (Section 26), the large toe (Sections 26 and 30), the teeth (Section 26), and the stump of an amputee (Section 30). All of these locations proved to be surprisingly effective, even though none had likely received much specific practice on the materials presented. Katz distinguished the vibratory sense of the skin, which serves to discriminate fine texture and roughness, from the spatial sense of the skin (pressure receptors), which serves to detect larger bumps. He covers touching at a distance (Sections 24 and 25), and makes the interesting suggestion that all touching is remote touching, given the horny layer of the epidermis, which is like having a layer of dried collodion on the fingertip. Similarly, Gibson (1966) said that touching may be done with nails, claws, hooves, and horns, in which case the stimulus is "not a direct impression on the skin by an object, as we tend to assume" (p. 100).

Finally, Katz discounts the role of felt resistance in the perception of roughness (Section 23), concluding from his tests that the roughness impression depends only on the coefficient of friction between the skin and surface, not on which one actually moved. However, Taylor and Lederman (1975) found that felt roughness depends mainly on the spacing of elements, not the coefficient of friction (see also Lederman, 1982). M. A. Heller (personal communication, October 1988) said: "I have noticed that adding a liquid to a surface (e.g., plain water) can cause qualitative changes in perceived texture, but not induce any 'quantitative' change. I suspect that magnitude estimation would miss these changes." Heller also said that fine (400-grit) sandpaper feels soft to some people; it almost feels cloth-like to him. Katz gives more attention to roughness than to hardness, but both dimensions are important, and their relationship may well be a fertile area for further study.

DIVISION II: QUANTITATIVE STUDIES OF THE TACTUAL PERFORMANCE

Chapter I:
Studies on Surface Touch

Part I: Experiments on the Qualities (*Modifikationen*) of Surface Touch

Section 18. The Tactual Material

First a general caveat: The research reported in Division II will not and cannot be exhaustive. I freely confess that the almost limitless possibilities for varying the experimental conditions has occasionally had a frankly inhibiting effect on my disposition to carry out the studies. Where should one begin, when such rich abundance beckons everywhere! I dare not conclude that the experiments I finally selected were the correct choices; someone else would perhaps have considered others more important. However, I now categorically refuse to indicate everything that could still be undertaken in this area. My experience indicates that providing that type of hint—sometimes kept mysterious to boot—in a footnote inhibits other researchers from taking up the problem.

To set some bounds, the investigation was limited from the beginning to a sufficiently large but not too extensive set of tactual material. All experiments in Part I were carried out with 14 different papers. Obtaining the materials was done somewhat differently than is usual in quantitative studies in psychology. To wit, the tactual stimuli could not be suitably produced for our experimental purposes with precisely stepwise differences (with an exception mentioned below in Section 24, Subsection 4), but rather a suitable selection had to be made from the hundreds of papers normally available. Regrettably, the clarity of the procedure suffers somewhat due to this. I cannot even begin to characterize all of the papers in the sense of commercial product specifications. Verification of the findings is thereby made somewhat more difficult, but not impossible.

I will first describe the series of 14 papers: 1) very smooth, very well waxed paper; 2) smooth, well waxed paper; 3) slightly waxed paper with a slight grain; 4) very fine grained writing paper; 5) fine grained writing paper; 6) smooth paper with a distinct grain; 7) soft paper with a slight roughness; 8) harder, not too rough, drawing paper; 9) soft blotting paper; 10) moderately hard blotting paper; 11) hard grained blotting paper; 12) hard, rough packing paper; 13) very hard, very rough packing paper; 14) soft, extremely rough cloth paper. (The cloth paper was that deep, black type frequently used for visual designs. As is well known, it consists of an approximately .5-mm thick layer of cloth fibers glued on paper. It might be objected here that we are no longer dealing with paper. The series was selected, however, without regard to the chemical-physical properties of the raw materials used, from materials customarily thought of as paper based on their visual and haptic qualities.)

I selected the papers so that by concentrating my attention and touching as I wished, I could distinguish each paper from every other one in the series. Since it was not possible to get around this subjective procedure in devising the series, usable data were assured in the following experiments only from observers who did not deviate too far above or below the experimenter in their tactual sensitivity. The ordering of the papers was such that the farther apart they were on the scale, the more their tactual impression differed in general. Thus, in most cases, adjacent papers had the greatest similarity with each other. The subjects kept their eyes closed when touching. In addition, their ears were stuffed with cotton after it became evident that noises originating from touching could give useful hints for recognizing the properties of the tactual material. However, once alerted to this source of error, many observers said they were not certain that they did not still have auditory impressions of many tactual surfaces in spite of their plugged ears. We will pursue this remarkable statement futher in a different context (Section 39).

Performance naturally fluctuated from person to person in these experiments, as is typical, but for now we must be less concerned with working out the individual differences than determining the general tendencies. The satisfactory agreement in the results presented below indicates that the performance obtained was average or typical.

Section 19. Basic Experiment

The papers on our scale will always be designated below by their numbers (1-14). They were presented on a pasteboard backing, and covered with a piece of cardboard having a rectangular opening, whose prominent edges indicated the boundaries of the paper to the touching hand. The papers were 10 by 15 cm wide, which offered the touch organ wide room for maneuver. In each instance, two papers were presented for comparison, as a rule, in the combinations 1 and 2, 2 and 3, 3 and 4, etc. Thus, 13 pairs were judged. If two adjacent papers on the scale could not be distinguished by the subject, then pairs of papers were presented which differed by 1, 2, 3, or more intervening papers. Insofar as no finer precision was obtained, we will consider as correct a judgment in which *two adjacent* papers were recognized as different, and as false any judgment which deviates from that. The subjects were instructed to behave as if they had to purchase papers in a store for an important practical purpose, while having to distinguish between the presented papers exclusively by touch. They were free to choose the hand, the fingers of the hand, and the way in which to touch. The temporal order was not prescribed, so the observers could begin with the paper surface lying at the left or right, and they could also return, if it seemed convenient, to the surface which they had touched first. The subjects were kept in the dark in every respect, i.e., they were told nothing about the purpose or the results of the experiment. Though no duration of touching was prescribed, the subjects did not abuse this freedom, but decided relatively rapidly. This confirms what I myself have observed time and again, which is that one has a remarkably fine feeling for whether or not longer touching would aid discrimination or recognition. The subjects were Miss Lommatzsch (L), a philosophy student; Dr. Keller (K); and Mr. Herbers (H), a theology student. L was lefthanded and used her left hand for touching, whereas K and H were righthanded and, as would be expected, touched with their right hands.

Under the experimental conditions as described, L and K correctly distinguished all papers from each other, and H distinguished all except 4, 5, and 6, which appeared the same to him. Although the instructions demanded only the identification of sameness or difference, the subjects often also provided qualitative judgments by speaking of writing paper, blotting paper, drawing paper, etc. I will disregard such judgments for now; experiments will be presented below (Section 23) in which qualitative judgments of the individual papers were requested in the instructions. Some papers feel cool and others warm; later experiments (Section 36) will reveal the interesting role that temperature plays in the recognition of materials. From the results obtained in the basic experiment, it appears that our selection of papers was appropriate; it was intended that all papers would be discriminable from each other under favorable conditions for touching. Essentially the same results were obtained in the basic experiment when the papers of each pair were not touched successively by the same hand, but simultaneously by both hands.

The report below of variations of the basic experiment will be interrupted here and there in order to point out more general findings of the experiments. Accordingly, we will now make a few remarks concerning the touching movement in the basic experiment, which will somewhat sharpen what was said above (Section 17). It is surprising how much difficulty the subjects had in answering when asked which fingers and what method they used in touching. Obtain for yourself a tactual impression from a piece of paper and surrender completely to it. You then will be able to confirm how modest is the degree of consciousness of everything that refers to the movement itself. With which fingers and which portions of the fingers did you touch? How long did you touch until the specific impression of paper occurred? With what speed was the movement carried out as a whole or in parts? To obtain information concerning these and yet other aspects of the movement, one would have to repeat the touching movement, and let the tactual impression itself recede completely into the background in order to make conscious the component contents of movement. Once the specific tactual impression of a paper has been established, it remains completely constant during further touching; it stands out quite sharply from the somewhat chaotic background fashioned by the touching movement. The basic experiment has aided us here in stating questions whose experimental answers will follow below.

Section 20. Variation in the SIze of the Tactual Surfaces

In the basic experiment, the tactual surfaces were so large that the observer never felt the need to touch beyond their boundaries. The first variation of the basic experiment involved reducing the size of the tactual surfaces. From the 14 papers on the scale, two sets of disks having diameters of 2 and 4 mm, respectively, were punched out, and each disk was pasted by itself onto a cardboard surface. Then, following the same procedure as in the basic experiment, the disks of each set were presented for touching. Subjects were L and K. L touched with the middle finger of her left hand, which she considered the most sensitive; K used his right index finger. The disks of paper were felt as low elevations of varying height on the background of the cardboard; the experimenter placed the touching fingers onto the correct position, i.e., laid them on the disks.

All of the 4-mm wide tactual surfaces were distinguished from each other by L, just as in the basic experiment, except that 8 was judged the same as 9, and 10 the same as 11. Thus, compared with 13 correct judgments in the basic experiment, there were 11 correct and 2 incorrect judgments here. Whereas most of the larger disks could still be recognized as consisting of paper, this no longer was the case with the smaller disks. Despite this, however, the differences between the individual tactual surfaces could still be distinguished remarkably well. L gave 10 correct and 3 incorrect judgments. K did little worse with the 2-mm wide disks: 8 correct and 5 incorrect judgments. It was most surprising that differences between such small touching surfaces could still be so reliably discerned. Punching out disks of 2-mm diameter produces a somewhat bent-down edge, which may be rougher or smoother than the paper itself, depending upon the type of paper, but as a rule is smoother. If subjects let their judgments be influenced by that edge as well, then that alone may explain the increase in erroneous judgments from the basic experiment. (This source of error cannot be eliminated by placing cardboards with openings on the tactual surfaces, and touching the latter through the openings. The touch organ no longer would come into contact with the tactual surfaces in the desired manner.) From this variation of the basic experiment, we establish that even a vast reduction in the size of the tactual surface impairs its discriminability only slightly.

A few spontaneous comments concerning size differences between the small disks seem worth mentioning here. To both L and K, for example, 12 appeared significantly smaller than both 11 and 13. These judgments probably are related to the perceived thickness of the tactual surfaces, with the thicker one appearing to be larger. Otherwise, the difference in paper thickness was unimportant for the results of the experiments. A quite remarkable size illusion occurred for Miss Kretzer (Kr), a philosophy student who occasionally participated in these experiments. Like K, she touched with her right index finger. She reported that the larger disks appeared to have a diameter of 3 to 4 cm. If she had given it some thought, she would not have doubted for an instant that the disks were very much narrower than her touching finger and therefore had to be considerably narrower than 3 to 4 cm. Considerable overestimations of the size of the disks also were typical for the other subjects.

What purpose was served by the variation of the basic experiment described in these paragraphs? If the eye is offered bits and pieces of various types of paper that are smaller than the smallest formal elements characteristic of them, then the unaided eye no longer can recognize the particular type of paper. One then sees, so to speak, a paper that is not defined by its more detailed constitution. Even under favorable conditions of distance and illumination, my nearsighted vision can no longer distinguish between small pieces of assorted paper if they cover an area of only about .1 mm^2. To be sure, I recognize that all of the little pieces are paper, but no longer what type of paper. The impression of an indeterminate paper disappears when the little pieces are even smaller. Then the impression of "some material" replaces that specifying paper. These visual observations provided the impetus for studying the effect of reduction in size of the tactual surfaces. As already reported above, most 4-mm wide disks were still recognized as being paper; going to 2-mm wide tactual surfaces completely destroyed this impression. But this step did not also eliminate all of the distinctions in tactual impression that existed between the different papers. One can expect that if the size were further reduced, a limit would soon be reached at which the papers could no longer be distinguished from each other. Experimental proof could not be obtained for this, because it was not possible to punch or cut out smaller tactual surfaces without getting salient and misleading irregularities at the edge. The next variation of the basic experiment permits this goal to be reached in a different way. Before we report it, it seems best to introduce the concept of reduction of tactual impressions, which will be useful in developing the investigation.

Section 21. The Reduction of Tactual Impressions

In the World of Color (Katz, 1911, Sections 4 and 8), the concept of reduction of color impressions was introduced, whereby complete reduction was distinguished from partial reduction (which, to be sure, was not yet so designated there). At both levels of reduction, a relatively slight variation of the effective conditions of stimulation makes the impressions caused by the color stimuli more similar to each other. If partial reduction makes the perceived differences in the illumination of the surface colors recede or disappear, then complete reduction gives all colors the character of film color, even if they initially had a very different mode of appearance. Reduction reveals itself to be, on the one hand, a methodological principle, a causal-genetic question, such as that involved in ferreting out the structure of color perception, and, on the other hand, a classification scheme for the simplification and unification of the color impressions. Complete reduction funnels all color impressions into that of film color, which on more than one ground is held to be a color phenomenon of a particularly simple structure.

The concept of reduction procedures will now be introduced for the present area of investigation as well. We will indicate methods quite analogous to those in the case of color, by which a relatively slight variation of the effective conditions of tactual stimulation produces a relatively weak or strong simplification and unification of the tactual impressions. Given the dissimilarity in type of stimulation (and the absence as well of anything analogous to the factor of illumination in touch), the procedures effecting a reduction in the case of tactual impressions are naturally completely different in nature than those in the case of color. Nevertheless, it still proves appropriate to speak of different steps in the reduction of tactual impressions.

In the experiments described in the preceding section (Section 20), the 2-mm wide disks could no longer be recognized as paper surfaces; according to the reports of the observers, something like wood, woven material, leather, and the like could just as easily have been involved. Cases where the experimental procedure destroyed the specific impression, thereby leading to an assimilation of different identifying characteristics (*Spezifikationen*) of surface touch, we will call a *reduction of the first degree*. We then speak of a *reduction of the second degree* when the procedure used also eliminated differences in qualities (*Modifikationen*), such as roughness and softness, which modify the identifying characteristics. One could distinguish even finer steps within the two degrees of reduction that we distinguished. The two degrees of reduction shade into each other in a manner analogous to that of the partial and complete reduction of color, namely, as the means of reduction are increased, a reduction of the second degree emerges out of that of the first degree. The stronger the means of reduction, the more the objective pole of tactual phenomena recedes relative to the subjective pole.

Whereas in the preceding section (Section 20), reduction was obtained by reducing the size of the touching surfaces, that is, by means of a change in the objective experimental conditions, the next section (Section 22) reports on reduction obtained by suppressing lateral tactual movement, that is, by means of a change in the subjective experimental conditions. The theoretically-instructive reduction procedures are predominantly subjective in nature, and our study is aimed primarily at them. A few of these procedures are mentioned here. Fatigue of the touch organ, e.g., as a result of heavy rubbing,[1] has a reducing effect. A reduction is effected by cooling,[2] by anesthetizing, and by diminishing the blood circulation,[3] the touch surface, or the pressure, as well as by excluding certain touching movements. The briefer the tactual contact, the more marked the reduction.

Unless otherwise stated, all of the following experiments were carried out at a comfortable room temperature and with the touch organ at a normal temperature.

Footnotes

1. Light rubbing has a restorative effect, presumably due to the increased blood circulation and the resulting improvement in nourishment of the touch organs. "The feeling of touch [in the blind] which is fatigued by prolonged gliding over the letters, is immediately refreshed again when one moistens one's fingers somewhat and rubs on a solid base a few times; even unclear writing thereby improves in clarity." E. Javal, *Der Blinde und seine Welt* (The blind and their world). Hamburg, 1904.

2. "As is well known, experiments on the sense of pressure . . . turn out poorly in a cold room. If one places a 5-deg C [constant-temperature Thunberg] temperator for 1 to 2 minutes on an area of skin where the thresholds of the pressure points are being investigated, then immediately afterwards a manyfold elevation of the thresholds over their normal values can reliably be established." M. von Frey, Die Webersche Taeuschung oder die scheinbare Schwere kalter Gewichte (Weber's illusion: The apparent weight of cold objects). *Zeitschr. f. Biol.*, 66, p 421. Editor: Green, Lederman, and Stevens (1979) found a sizeable decrease in the perceived roughness of grooved plates, especially on plates with grooved widths of less than 0.5 mm, as skin temperature decreased from 45 deg C to 10 deg C. Apparent roughness declined as skin temperature

fell below normal (32 deg C), but was enhanced as skin temperature rose above normal.

3. "The effect of hyperemia [superabundance or congestion of blood] and anemia [deficiency of blood] on the individual sensory qualities is essentially that hyperemia decreases the thresholds and anemia increases them." F. Hacker, Versuche ueber die Schichtung der Nervenenden in der Haut (Experiments on the layering of nerve endings in the skin). *Zeitschr. f. Biol., 64,* p. 218. Fanny Halpern found that increasing the blood circulation first increased and then decreased tactual sensitivity. Ueber die Beeinflussung der Tastschwelle durch aktive Hyperaemie (Influence of active hyperemia on the touch threshold). *Pfluegers Archiv, 197,* 1922.

Section 22. Touching without Lateral Movements on the Tactual Surface

Subjects L and Kr touched the papers of the basic experiment with the touching finger used in the experiments described in Section 20. They were instructed to approach the tactual surface from above, but *carefully* avoiding lateral movements on it. They were permitted to repeatedly raise and lower the finger on the tactual area in this way.

Subject L no longer recognized that she was dealing with paper, and in addition almost all of the differences she had perceived between the 14 papers in the basic experiment had disappeared. Only Paper 14, by its softness, could be distinguished from all of the other 13 papers. Subject Kr did nearly as poorly except that she detected differences in hardness or softness between a few papers that L missed. Thus, a reduction of the second degree, which could not be obtained by an extensive reduction in size of the tactual surface (Section 20), was nearly achieved here by the suppression of lateral touching movements. What now remains here as a tactual phenomenon? A surface touch about whose specific qualities no sure statement can be made.

Earlier, in discussing the identifying characteristics (*Spezifikationen*) of surface touch, the roughness-smoothness and hardness-softness dimensions were set side by side as equivalent. We now add that that was not entirely correct, for they are differentiated by reduction, inasmuch as roughness and smoothness arise during lateral touching movements on the surface, whereas hardness and smoothness are revealed in the movement of the touch organ vertical to the tactual surface and depend on the depth dimension. It is surprising how many tactual perceptions are provided by temporal, spatial, and intensity variation of the excitation of the fingertip produced by movement vertical to the tactual surface. Such movement can provide information not only about all degrees of hardness and softness, but also all of the degrees of elasticity that cardboard, metal surfaces, stretched rubber sheets, and other elastic materials possess. More about this later.

Given that almost all differences between the papers disappear when movement is restricted to that vertical to the tactual surfaces, with no lateral movement allowed, such differences ought to be even less available to the motionless finger. However, the latter deduction is apparently contradicted by the following reports made by all of the observers in the basic experiment as well as by another, special observer. If one of the 14 papers has been recognized in its particulars during a customary touching movement and then the finger comes to a complete halt, the impression of that paper persists with complete clarity for a relatively long duration. The following procedure, whose results were already mentioned above (Section 12), produces a particularly convincing gestalt. Place before the subject two papers whose difference is detected instantly if they are touched simultaneously (one with a finger of the left hand, and the other with a finger of the right), but completely missed when all lateral motion in those fingers is excluded. If, after abundant movement, both fingers simultaneously come to a halt, then tactual images persist which can be easily distinguished as such. No one would suppose that the direct stimulus excitation of a *motionless* tactual surface differs depending upon whether the touch organ came to rest in this way or that. The only remaining possibility is that we are dealing here with purely subjective tactual images, about whose nature an alternative was already stated above. Less differentiated and less persistent tactual afterimages occurred in the experiments with tactual movement vertical to the tactual surface. The touch phenomenon experienced when one's finger comes to rest on a tactual surface following a downward movement is quite definitely influenced for the first 1 to 2 seconds by the persisting tactual image, which develops as the fingertip moves from its first collision with the paper to a state of rest.

Section 23. Experiments with Movement of the Tactual Surfaces

1. *Experimental method.* The friction that arises between two moving surfaces depends only upon the movement of the surfaces relative to each other, not upon how the movement is distributed by amount and direction between the two surfaces. If the tactual impression of a surface depends only upon the magnitude of the friction between it and the touch organ, then it should not matter whether the touch organ moves over a motionless tactual surface, or, conversely, the tactual surface moves at the same speed over a motionless touch organ.[1] Even if the quotation from Weber presented above (Section 14) did not exist, hardly any researcher acquainted with the

numerous paradoxes of perception would want to doggedly defend the impossibility of such a difference. Whoever associates the degree of distinctiveness of roughness and smoothness with the degree of frictional impediment to the actively moving touch organ, must expect that transferring the movement from the touch organ to the tactual surface will profoundly influence the tactual impression. However, we will do without any such theoretical construction, and will let the experiment decide the matter.

In order to move tactual surfaces past the completely motionless touch organ at a selected speed, the following apparatus was constructed (Figure 1). A Schumann's tachistoscope was fastened to a table in a horizontal position. At its edge, a sheet-metal support for the tactual surfaces was fastened so that it was centered approximately 50 cm from the axis of rotation. The support moved about 3 cm below a very stable piece of cardboard on which the subject laid his or her hand. The subject was comfortably seated, and placed the hand so that its fingers, except for the thumb, extended beyond the cardboard. The fingers were usually held extended, and only when the experimenter called out, shortly before the tactual surface in question moved below, were the fingers bent down so as to make good contact with the tactual surface. As soon as the contact was made, the fingers were to be kept still, that is, they were not allowed to execute any active touching movements. By contrast, in the basic experiment, the type of touching movement was left completely up to the subjects. He or she could use this or that finger, could move to and fro more or less frequently at a particular speed, could exert more or less pressure on the tactual surface. Such freedom also was allowed here. Preliminary tapping experiments showed that the tactual impression with a motionless touch organ depends very considerably upon the pressure exerted on the tactual surface; the surfaces all appear smoother with weaker pressure than with stronger pressure. These preliminary results led us to instruct subjects to exert a moderate pressure on the tactual surfaces or to let the tactual surfaces glide by under the fingers at a moderate pressure. This may have made the surfaces appear somewhat smoother than when the touch organ was actively moving. The papers to be touched were fastened on the support with clamps so that they could be easily interchanged. A rather thick strip of cardboard was placed over the front edge of the support to inform the subject of the immediately following tactual surface. The tachistoscope was moved by a motor, whose speed could be varied within wide limits by means of a rheostat.

The comparison procedure of the basic experiment was not employed here. Instead, subjects made qualitative judgments of each individual paper on our 14-member scale. Thus, as is true of any absolute judgment, more was demanded in this case than in the basic experiment. In order to compare the performance here with that in the basic experiment, the latter was repeated in such a way that the 14 papers were judged qualitatively. The pos-

QUANTITATIVE STUDIES OF THE TACTUAL PERFORMANCE   103

Figure 1.

sible influence of practice, due to repeating the basic experiment in a new form, need not be considered further, because, if anything, it aided performance to an even greater extent in the subsequent experiments using the tachistoscope. The two experimental configurations being compared thus were not entirely equivalent.

The first experiments were carried out with L and K. First, here are their qualitative judgments in *repeating the basic experiment.* L's statements: 1. Very smooth, glossy(!). Could be smoothly polished metal. 2. Very smooth, like lacquered metal. 3. Smooth, nonglossy paper. 4. Very smooth paper, not for writing. 5. Heavy, smooth writing paper. 6. Very light, smooth paper, not shining. 7. Very light paper as in writing pads, very smooth. 8. Drawing paper. 9. Very fine blotting paper, rough, fibrous. 10. A type of drawing paper, very flat. 11. Drawing paper, medium heavy, not very rough. 12. Slightly granular paper with surface irregularities, not fibrous. 13. Rough, nonfibrous, granular, heavy drawing paper. 14. Felt-like paper, rough with little hairs. K's statements: 1. Very smooth paper, like porcelain or still better, because of the small humps, like stoneware. 2. Also very smooth paper, but not as smooth as the preceding. It is probably glossy paper. Imagined as being shiny visually. 3. Very smooth paper, but doubtful whether it is glossy paper. One could hardly write on it with a pencil, because the writing would not adhere sufficiently. A pointed pen would probably "run it through." Imagined as having a matte gloss. 4. Not a glossy paper, but a highly smoothed writing paper for lead pencil and ink. 5. Good for writing with pencil, not as rough as newsprint. 6. Rougher than the preceding paper, but not newsprint. Not suitable for ink because the pen would be caught on the irregularities. 7. Smoother, medium grade paper similar to note pads. Ink could spread out on it. 8. Very rough, good tough paper, very well suited for pencil, one would get stuck on it with a pen. 9. Fabric-like paper. Suitable for pencil, not for ink. Similar to crayon paper. 10. Even coarser and rougher, woollier, and more porous than the preceding paper. Not suitable for ink. Similar to blotting paper, but not as soft. 11. Very coarse paper, actually woody, perhaps packing paper. 12. Very woody portions, on which one is caught. Perhaps packing paper. 13. Even somewhat rougher than the preceding paper, but does not have such markedly spike-like points. 14. Fabric-like as if black cloth.[2]

Reproducing the reports in such detail was unavoidable. However, I believe that this completeness is also quite well suited to providing the reader with a formative image of the actual tactual performance. K tended to draw comparisons between the successive tactual impressions in characterizing them, even though only absolute judgments had been requested. K was not alone in this regard; we also observed it in other subjects who served in these experiments. I have already drawn attention here to the frequent occurrence of visual images. We may presume that they had been numerous even where not expressly stated. However, the significance of the concomitant visual images will not be dealt with in more detail until later (Section 46). We will not discuss how the tactual performance of the basic experiment compares with that obtained here. The question would not be very simple to resolve, in any case, because the difference in instructions alone could have produced a difference in mental set and thereby in performance. We limit ourselves to a comparison of the performance with a moving touch organ and that with a moving tactual surface; in both cases, the instructions required qualitative judgment of the individual tactual surfaces.

2. *Experiments with medium speed of movement of the tactual surfaces.* We now describe the experiments with a moving tactual surface. The motor was set so as to provide the central portion of the support surface of the tachistoscope with a speed of approximately 15 cm/sec, which appeared appropriate based on our pilot experiments. The papers were presented (only once each) in random order, but we reproduce the statements concerning them in proper order. The subject was left completely in the dark as to whether the same papers were used as in the preceding experiment. Here, too, in addition to the individual qualitative judgments, there were comparisons made between the many papers, which even involved reference to the preceding experiment with the moving touch organ. It will suffice to report in detail K's judgments: 1. Extraordinarily smooth paper, was probably the smoothest of the preceding experiment (i.e., the experiment with a moving touch organ). 2. Smooth, but not as smooth as 1 (1 preceded 2 due to the random ordering). 3. Fairly smooth, almost like glossy paper, but not that of the preceding experiment. 4. A very smooth paper, but not glossy paper. 5. Smooth writing paper. 6. Smooth paper similar to 7 (7 preceded 6

in the experiment, so the subject could refer to it). 7. Moderately smooth, probably thin, smooth writing paper, through which one can feel the base. 8. Moderately rough, ordinary writing paper. 9. Moderately rough, somewhat coarser writing paper. 10. Relatively rough drawing paper. 11. Rough, heavy drawing paper, not woody. 12. Very coarse, woody paper, not suitable for writing. 13. Very coarse paper, not suitable for ink. 14. Cloth paper.

Two additional observers, who had not participated in the previous experiments, were brought in: Mr. Zerck (Z), and Miss Ehrenberg (E), both philosophy students. The experimental conditions were exactly the same as with L and K. To save space, I will present only Z's reports. *Moving touch organ*: 1. Quite smooth paper, glossy paper, not for writing. 2. Not as smooth as 1, but also smooth. Not suitable for ink. 3. Not as smooth as 1 and 2, suitable for writing. 4. Smoother than 3. 5. A poor quality, rough writing paper. 6. Smoother, but also more granular than 5, not suitable for ink, perhaps for pencil. 7. Smoother and finer grained than 6, but not suitable for writing. 8. Solid, thick, rough paper, perhaps so-called linen paper. 9. Rough, fabric-like paper, finely fibrous, not suitable for writing. 10. Fabric-like paper, rougher and thinner than 9. 11. Rough paper, firmer than 10, suitable for softer pencil. 12. Inferior paper, with smooth and rough portions, not suitable for writing. 13. Coarsely grained, non-fibrous drawing paper. 14. Fabric paper. It is striking that almost all of Z's judgments also have a comparative character, even though, I repeat, the instructions themselves did not suggest comparison. Judgments 4 and 7, inasmuch as they refer to preceding papers, are not pertinent, but otherwise we can designate the qualitative judgments as satisfactory or even good. Here are Z's reports with a *moving tactual surface*: 1. Completely smooth, glossy paper, as smooth as 1 in the preceding experiment, but not as soft. 2. Smooth, non-glossy paper. 3. Smooth paper, but suitable for writing. 4. Smooth, almost glossy paper, too smooth for writing. 5. Smooth, but not completely flat writing paper. 6. Smooth paper, but bumpy rather than flat. 7. Smooth, like paper from writing pads. 8. Rough drawing paper for pencil. 9. Fine grained drawing paper. 10. Fine grained like 11 (which immediately preceded it in the experiment), but rougher and not suitable for writing. 11. Fine grained paper, for writing with a broader pen. 12. More coarsely grained and rougher than 9 (which immediately preceded it in the experiment), but not suitable for writing. 13. Quite rough, thin packing paper. 14. Fabric paper, appears firmer than when the surface is motionless.

It would be too much to contrast the two experimental configurations by comparing each subject's two judgments of each individual paper, and presenting the detailed results here. That task, insofar as it can be done based on the material reported for K and Z, must be left to the reader. Here we can only render a summary impression of those comparisons. At the speed of tactual surfaces that we selected, the tactual performance of all four subjects fell somewhat, but not too far, behind the performance obtained when the touch organ was moving. In order to properly evaluate this result, we must consider the points of difference in the two configurations. How does the outward difference in the two types of movement work itself out in detail? Tactual performance must suffer if the tactual surface moves past at a speed of 15 cm/sec, so that its relevant, 20-cm length acts as a stimulus only for about 1.25 sec (and is moved past only once), whereas with the moving touch organ, the subject is left free to decide how long to continue with the touching. Greater advantage of this freedom naturally was taken, in particular, in the more difficult cases, but even in the easier cases, no doubt, the 1.25-sec period was always exceeded. Research on tactual recognition time is presented below (Section 29). The motionless tactual surfaces also had an advantage in that all five fingers of the hand could participate in the touching and, in fact, usually did so, while in the experiments with a motionless hand, the thumb was completely excluded and probably the little finger, too, for the most part. As we already know from everyday experience, five fingers touch better than three or four, and subsequent experiments also will confirm this. The moving touch organ had a further advantage in that the subject could favor those portions of the fingers that promised to be most effective, while it was more fortuitous as to whether those portions also would be used when the position of the hand was fixed once and for all. And finally, the configuration with the moving touch organ was favored in that the touching movements could be carried out at an adjustable speed, which was left up to the subject, while in the other configuration the tactual surfaces moved past the motionless hand at a fixed speed. Subsequent experiments (Section 33) will show that each subject has a particular speed which gives the best tactual performance. Spontaneous remarks by the observers confirmed all of these differences between the two configurations, except for the last point.

Since the factors just listed, which indicate the undue advantages accorded to the moving touch organ, can explain the differences in performance, we may conclude that it matters little whether the movement in touch is exclusively in the touch organ or exclusively in the tactual surface.

Thus, what is decisive for the recognition of the identifying characteristics (*Spezifikationen*) of surface touch is the physical emplacement of friction between the touch organ and the tactual surface. This contradicts the opinion of those authors[3] who consider the experience of greater or lesser effort resulting from *active* movement of our touch organs over the tactual surface to be the basis of our judgments of smoothness and roughness. Our experiments with a motionless hand could not have involved such an experience of effort, yet the tactual performance did not thereby noticeably decrease. I wish to add here yet another counterinstance. I pasted Paper 1 onto one side of a thick cardboard, and Paper 14 on the other side, and then touched both sides simultaneously with the fingers of one hand, the thumb on one side, and the index and middle fingers on the other side. I clearly had both tactual experiences next to each other. The experience of resistance in the wrist is thereby shown definitely to be irrelevant, since it cannot lead simultaneously to two completely different judgments of roughness. One could perhaps modify the theory just dismissed, by declaring that the experience of roughness is based on stretch sensations in the skin or in the joints that still would be activated even when the touch organ is kept passive. In no way do I deny that something like that occurs. With passive touching, differences in the strain sensed at the skin and in the loading of the joints of the touch organ may reveal relatively large differences in roughness, but not, I believe, relatively fine differences. But even where such well-based experiences of resistance implicate properties of tactual surfaces, the function that connects the two dimensions with each other is not always so simple. Here are two examples of this.

When a warm hand is moved over a metal or glass surface, then the perspiration that occurs can fundamentally change the frictional conditions. Whereas the hand moves continuously when it is dry, it now proceeds jerkily, the fingers stumble somewhat across the surface as they now catch and now slide and jump (see also the Addendum of Section 36). Stronger experiences of resistance, which can halt the touching movement, do not necessarily make us judge that the metal or glass plate is rough.[4] Paper 14 lies before me. I first touch it with one finger, then with all five fingers and the palm simultaneously. The paper feels equally rough in both cases, yet undoubtedly much more force had to be applied in the second case in order to overcome the frictional resistance. If the applied force alone were the cue for judging roughness, then I could not arrive at the same judgment for the paper in both cases. One would have to make the additional assumption that the surface area of the touching organ is taken into account in making the judgment. To a limited extent that may be true, but certainly not in such a sensitive manner as the recognition of finer differences in roughness would demand.

I have not systematically studied the extent to which a motionless touch organ impairs recognition of the identifying characteristics (*Spezifikationen*) of surface touch, but I should like to venture the hunch, based on occasional observations, that the impairment lies within the limits shown above by our experiments on the qualities (*Modifikationen*). Weber's conclusion in the quotation noted above (Section 14) leads to a different expectation. An explanation of this divergence is not possible, however, since the experimental conditions reported by Weber are not completely clear.

3. *Experiments with variation in the speed of movement of the tactual surfaces.* A systematic investigation of the influence of the speed of movement of the tactual surfaces on the touch phenomena was not my intention, but some

tests were conducted before the experiment described above in order to find the most suitable speed for the main experiment. A few more words are in order about this. I found that a medium speed of about 15 cm/sec brought out most clearly the differences between the papers. Beyond a speed of approximately 60 cm/sec, all papers felt much smoother than when touched normally. At that speed, the differences in roughness are not as well perceived, but rather all the papers become more similar to each other without becoming completely identical. Below a certain speed (about 3 cm/sec), all of the smoother papers feel rougher than when touched normally, and at the same time all the papers become somewhat more similar without becoming completely identical. Thus, we find that above and below certain limits on speed, movement makes matters worse (reduction). We will speak about these experiments again in a theoretical connection.

Footnotes

1. As far as the perception of movement itself in both cases is concerned, Mach (op. cit., p. 152) has already given attention to this question: "We know quite well how to distinguish between passing our fingertips over a motionless object, and the movement of an object over our motionless fingertips. Also relevant here are the analogous paradoxical phenomena connected with rotary vertigo. They were known to Purkinje." According to A. Basler, "the threshold for the movement sensitivity of the skin fluctuates at its lower level at the speed that can be recognized by the eye at a distance of 30 cm." Ueber das Erkennen von Bewegung mittelst des Tastgefuehls (Recognition of movement by means of touch). *Pfluegers Archiv, 136,* 1910, p. 384.
2. One can glean from occasional statements that the touched papers, as must be the case for other materials as well, differ in their degree of pleasantness. I have not pursued this aesthetic factor further. Soft, smooth surfaces can provide blind subjects with a high degree of aesthetic pleasure. K. Groos (*Die Spiele des Menschen* [Human games], 1899) mentions Richard Wagner's pleasure in touching satin, and Sacher-Masoch's joy in touching furs. Kurd Lasswitz (*Auf zwei Planeten* [On two planets]) has the Martians invent touching games. Helen Keller tells us of her aesthetic experiences in palpating classic statues. The handling of tactual surfaces by little children is well known to be very close to pure play.
3. Thus, for example, Schapp: "With a rough object, we hook onto every point and have to reduce the pressure in order to be able to feel along the surface. With a completely smooth object, however, we move easily and effortlessly over the surface (op. cit., p. 36)."
4. Physics distinguishes between adhesive friction and sliding friction. As a result of perspiration, the adhesive friction, which we could infer to a certain extent from the skin tension, is very great. In a touching movement, the two types of friction shade continuously into each other. For many tactual surfaces, finer

discrimination of roughness differences appears to occur with the hand than with physical science methods.

Section 24. Veiling Intermediaries

1. *Intermediaries that are tightly attached to the skin.* When I began investigating the recognition of surface structure, I presumed that the epidermal ridges found in their most consummate form at many points on the hand and fingertips would play a special role in these experiments.[1] This presumption was not fully confirmed, however, because portions of the body lacking such ridges showed good tactual performance, and, more importantly, the capability of the fingers was reduced less than expected when these ridges were made less effective. In any case, this presumption motivated the first experiment, which was carried out with veiling intermediaries.

The fingers used for touching in the basic experiment were immersed in a solution of collodion in ether. Evaporation of the solvent left the fingers coated with an approximately .1-mm thick collodion membrane, which completely smoothed over the differences in height of the papilla lines. The membrane was completely dry and showed not the slightest stickiness against touched surfaces. It sat absolutely fixed and could not be displaced on the skin. It was not noticeable at all as long as the person did not touch. How it manifested itself in touching will now be discussed.

After Subject L's fingers had been prepared as described, the basic experiment was repeated exactly. To my surprise, the touch performance was left nearly unimpaired. All of the papers were discriminated just as reliably as in the basic experiment, except that 5 was incorrectly judged as somewhat smoother than 4, and 7 as somewhat finer than 6. The judgments mainly referred to differences in roughness, which presumably were more readily accessible through the veiling medium, and referred less often even than in the basic experiment to differences in granularity. The basic experiment likewise was repeated with Subject Kr. Whereas all judgments in the basic experiment were correct, that is, she had correctly recognized the difference in all 13 cases, the veiling medium caused three incorrect judgments (4 smoother than 3, 5 rougher than 6, and 11 rougher than 12). Since the incorrect judgments were not the same for L and Kr, it is unlikely that they depended specifically upon the particular veiling medium used. The experiments show that tactual surfaces are not judged that much more poorly when they no longer come into direct frictional contact with the touching skin, but are separated by an immoveable intermediate medium. For the first time in our quantitative research, something like a remote effect was revealed.

The next experimental step was thus prescribed—go to a thicker intermediary. First, however, a few words are still in order on how the collodion membrane, which is not felt at all on the motionless touch organ, becomes noticeable by touching. The impression is not that of a touch-transparent film, as described earlier, but rather one of touching with fingers that have been superficially burned by a hot stove and thus have a smooth, glossy surface. Numb fingers in everyday life usually are those that have been exposed to severe cold for a long time. A certain connection exists between the tactual impressions obtained with collodion and with fingers cooled in that way. It is very important to note, however, that although the recognition of differences between the individual papers suffered little due to the veiling intermediary, the absolute impression obtained from each tactual surface was changed considerably. We will discuss such absolute changes below.

In the next experiment, the touching fingers were covered with adhesive tape, which was approximately .2-mm thick. Fresh tape is well known to stick quite tightly, allowing not the slightest movement with respect to the skin. Given the results of the experiment with collodion, it came as no surprise that the layer of adhesive tape hardly impaired the recognition of tactual surface differences. Subject L erred only by designating 7 as smoother than 6, and 10 and 11 as identical papers.

Certain visual experiments are highly analogous to experiments of this type on touch. When one looks through an achromatic (or colored) episcotister [which is a rotating disk with an open or cut-out sector], then the resulting color impressions depend upon the episcotister's color, illumination, and size of opening (Katz, 1911, Section 10). Nothing seen through the episcotister escapes its weakening and discoloring influence. But no matter how great the joint displacement of the color impressions of the visual field (borrowing from music, one could speak here of a color transposition), their relative displacement with respect to each other is slight. Differences in brightness, saturation, and hue between the individual portions of the visual field remain almost completely unchanged when viewed through the episcotister, and the practically important recognizability of the objects remains essentially intact. To be sure, a certain assimilation of the color impressions occurs due to the superimposed light of the episcotister, and the finest color differences are obliterated. But unless the episcotister transmits considerable light to the eye, the assimilation is so minor that it can only slightly disturb recognition.[2] Color thresholds and color matches can be determined almost as precisely through the episcotister as when it is removed. However, when two colors must be adjusted until they are identical or have some prescribed degree of difference, and one color is observed with and the other without an episcotister, then the task proves to be very difficult and it is resolved with nowhere near the perfection evident when both colors are transposed by the episcotister. The visual intermediary of the episcotister corresponds to the tactual intermediaries of collodion or adhesive tape in our touch experiments. Just as the episcotister produces a general color transposition, so, too, the tactual medium produces a general displacement of all tactual impressions. Just as the visual intermediary hardly impairs the recognizability of color differences, so, too, the tactual intermediary hardly conceals the tactual differences of the touch surfaces. This analogy leads us to predict that recognition of differences in the tactual surface must suffer greatly when one tactual surface is touched with and the other without a tactual intermediary. Experiments presented below fully confirm this prediction. The analogy aids us yet further in posing a completely different question. After the eye looks at a colored surface for a long time, it becomes fatigued for that particular color, and one generally speaks of an adaptation. The adapted eye then sees all colors in the visual field changed in a way similar to when the normal eye looks through an episcotister. Everything we have just said about vision through an episcotister applies as well to color vision with the adapted eye. The adaptation fades very rapidly, but indeed may still persist even after it no longer intrudes subjectively. Weak degrees of monocular adaptation can be demonstrated by means of binocular color matching. One can also adapt a touch organ. If a glove is worn for a long time and then removed, everything feels different and the hand appears to have become more sensitive. You can best convince yourself of that by comparing the impressions of this hand with those of the hand that had worn no glove. To be sure, there is no adaptation due to fatigue in this case. Experiments involving fatigue, which are presented below, will show that tactual adaptation fades very rapidly, much the same as visual adaptation.

2. *Intermediaries that lie loosely on the skin.* The fingers used in the basic experiment were covered with rubber fingers and the basic experiment was then repeated. Used first were rubber fingers of the thinnest type commercially available. The results were that L distinguished all papers except for 9 and 10. To be sure, in two cases the judgment of relative smoothness was the reverse of that in the basic experiment; 7 was termed clearly smoother than 6, and 12 clearly coarser than 13. A definite impairment of performance can be seen in the failure to distinguish 9 and 10, but not necessarily in the two cases of reversal, even though that may sound paradoxical at first. Suppose that the judgment of the properties of a tactual surface is determined predominantly by the frictional coefficient between the tactual surface and the touching surface, and that the epidermis covering the touch organ in one experimental group is replaced by a medium having a different frictional coefficient vis-á-vis the tactual surface. Then a reversal might very well occur in the judgment on the difference in roughness between two tactual surfaces, owing to a corresponding change in the frictional coefficients. The fact that all tactual surfaces appear qualitatively changed with the rub-

ber fingers, fully confirms the view developed here. As long as tactual surfaces are discriminated virtually as well with as without an intermediary, there is no danger of an error. Only an intermediary that swallowed up all differences would seriously mislead us. We conclude from these considerations and the results obtained above, that rubber fingers of the type used here do not significantly impair tactual performance compared with the basic experiment. The results were confirmed with Subject K. The judgments of differences in the tactual surfaces agreed with those of the basic experiment, except that 5 appeared rougher than 6, and 11 rougher than 12.

The thin rubber fingers were then replaced by very thick ones, approximately of the stiffness of kid leather. Subject L judged all papers as different, although the relative roughness of some papers was the reverse of that in the basic experiment. With 7 again appearing considerably smoother than 6, 10 finer than 9, and 12 finer than 11, this, too, may be explained in terms of changed frictional coefficients. In comparing 5 and 6, L felt less resistance in gliding over 5 and therefore concluded that it was smoother, although actually there was no apparent difference in the granularity of the two papers. I cite this to show that by no means need the degree of smoothness and granularity run in parallel. The last experiment indicates that even relatively thick intermediaries do not significantly reduce tactual performance compared with the uncovered touch organ. The experiments with rubber fingers also were repeated with the fingers of both hands covered and with one paper in each pair presented to the left hand and the other to the right. This variation of the experiment led to essentially the same results as the preceding experiments.

How were the rubber fingers experienced? In contrast to the case with the tight-fitting membrane, here the feeling of a touch-transparent film between finger and tactual surface persisted during all touching. I obtained comparisons of the impressions of fingers, in which one finger was covered with collodion and the other with rubber. Subject L: "The collodion mass belongs to the finger as if its skin were thicker or the finger had become duller after touching for a long time. With the rubber finger, by contrast, one has the impression of an intermediate layer not connected with the finger."

In medical practice, rubber fingers and gloves are occasionally used in the examination of internal organs as well as in surgery, when there is danger of infection. If infection could be prevented as easily by other means, then one would certainly do without the isolating, but confining, rubber membrane. To be sure, it is primarily stereognostic impressions that the physician attempts to obtain when examining the internal organs and their changes, or on which he or she relies when manipulating the organs, but even so the surface properties of the objects being examined have some influence on the stereognostic impression. In the physician's practice, it is almost exclusively mucous membranes that are felt, that is, tactual surfaces that differ completely from those we have investigated up to now, but here, too, touching via an intermediary ought to modify the tactual impression. The experiments presented below, in which surface touch with and without an intermediary is compared, may therefore also have a certain interest for medical practice.

Subject L was instructed to compare two papers with each other by touching one with the bare right middle finger and the other with the rubber-covered left middle finger. In one series of tests, pairs of identical papers were always presented at the left and right. In another series, as in the basic experiment, pairs of different papers were presented, and spatial location was varied as well. When the same paper was presented at the left and right, the identity of the two sheets was never confidently recognized, as is the rule when two bare fingers are used instead. Less simple are the results from the tests with different papers presented at the left and right. The following two cases occurred: 1. With both left-right spatial orderings of a pair, the result of the comparison was the same as in the basic experiment. 2. With one left-right spatial ordering, the two papers received the same judgment of difference as in the basic experiment, but with the other spatial ordering they were not discriminated or the judgment of difference was even reversed in sign. It is also worth mentioning that occasionally the same judgment of difference occurred as in the basic experiment, but with *greater* confidence. In summary, the results indicate that the comparison of tactual surfaces with unilateral intermediaries is very markedly impaired as compared with the case in which both sides are touched with bare touch organs or both sides are touched using an intermediary. Tests with Subject K confirmed this result. A practical rule that can be derived from these tests is that to compare two tactual surfaces, one of which *must* be touched with a covered touch organ, it is advisable to touch the other not with a bare touch organ, but rather using the same intermediary.

Considering what was said above about the far-reaching analogy between the experiments with visual and tactual intermediaries, which let us virtually predict the outcome of the last experiments, there is no need for a more detailed explanation at this point.

3. *Remote touching with a stiff intermediary.* Persons who have to write a great deal usually have a pronounced preference for a particular type of pen, which makes them pass up all other pens if at all possible. Differences in the friction of pens on paper, owing to the type of construction and the material of the pen, are so obvious that a real bureaucrat will not readily accept a new type of pen. As is more generally known, in writing with pencils one can discriminate at least a few degrees of hardness. Since everyday observations may be influenced by knowledge of the real conditions, I carried out a few

informal tests on this point with Subject K. Five pencils from the Johann Faber Company, with hardness values of 1 to 5, were to be used to write on the same paper, with the eyes shut and the ears stopped. The subject could discriminate quite well all degrees of hardness from each other. In the same vein as those oft-cited observations of Lotze (*Microcosmus*) on external projection, the hardness or softness of the pencil is localized at its point, not on the parts of the hand that guide the pencil, as long as one regards the matter naively. The tactual surface itself also develops at the pencil point. If the matter is regarded less naively, that is, more critically, then one notices in the grasping fingers themselves both a continuous pressure and a vibration that varies with the degree of hardness of the pencil and is generally more pronounced with hard pencils than soft ones. Subject K reported that the difference in hardness was more readily recognized in writing down strokes than up strokes. While writing, the subject had clear visual images of the writing strokes as thick and black, or thin and gray, depending upon whether the pencil was judged as being soft or hard.

Yet other experiences in writing deserve mention in this connection. Even when the same pen and the same paper are used, the impressions received while writing vary markedly with the type of pad on which the paper rests. One needs only to write one after the other on pads of cardboard, wood, leather, linoleum, glass, and metal, to discover that in each case one has a different experience. Naturally, care must be taken that the latter pads do not themselves have large differences in roughness that can be felt through the overlying paper. The differing elasticity of the pads apparently makes the pen vibrate in different ways (writing noises provide a definite indication that oscillatory processes are occurring here), and these vibrations then enter our consciousness. With prolonged writing, the unsuitability of an insufficiently elastic pad is unmistakably betrayed by the unusual degree to which we become fatigued. The elasticity of the writing pad does not display itself, like roughness and smoothness, at the surface of the paper written upon, but in depth, as we previously learned is the case for the hard-soft dimension. Furthermore, elasticity, in our view, must be recognized as a property of the tactual world that can be experienced directly and not only gauged in a physical science manner. Within very wide limits, the elasticity and hard-soft dimensions may be varied independently of each other. A thick woollen material is very soft, but not elastic; a piece of fishbone is very elastic, but not soft.

If we can discriminate the degree of hardness of various pencils while writing on the same paper, then this must be due to the frictional coefficient between the constant paper and the variable graphite mass. It should matter little or not at all, if instead we vary the papers while keeping the pencils constant, as long as we keep the same frictional coefficient. We have not carried out precisely this experiment, but have done a similar one. Our 14 papers were touched in the same way as in the basic experiment, using a wooden rod that did not project a very sharp point. A pen holder was used that was grasped as such; its point extended 4 to 5 cm beyond the fingertips. Besides Subject K, the observers were Mr. Noldt (N), Miss Dobschuetz (D), and Miss Raspe (R), all philosophy students. Subject K had 8 correct and 4 incorrect trials, and the other three subjects had 4, 4, and 5 false trials, respectively. The papers were no longer recognized as such. When questioned, subjects said that paper, but also other materials as well, could have been involved. But one must admit, nevertheless, that the fact that the surface type of most papers can be discriminated with the wooden rod is rather unexpected. It would be impossible for the pressure receptors in the skin to obtain even approximately corresponding pressure images, based on the minimal differences in level present in the paper surfaces. If the roughness discrimination was based solely on the difference in the effort experienced in moving the hand, then the result, though still remarkable in itself, would not lead much further theoretically. However, this is not the case. Differences in the experience of resistance undoubtedly play a role in these experiments, but only in detecting differences between papers that are far apart in the series. The finer differences of adjacent papers are recognized by differences in the vibrations that arise from rubbing the wooden rod against the paper and are transmitted by the rod to the hand. When asked, the subjects explicitly indicated the basis in experience for their judgments to be the vibration sensations, which need not be limited to the grasping fingers, but may also occur in diffuse form in other portions of the hand.

Remote touching by means of a rigid intermediary is effective only if the vibrations produced by friction are transmitted largely undamped to the touch organ. If the wooden rod is wrapped with a material that absorbs vibrations, e.g., felt or cloth, then most papers can no longer be discriminated. Papers lying far apart from each other in the series are still judged as different, but only because of differences in the resistance afforded to movement of the hand. It makes no difference whether the damping medium is applied between the hand and the wooden rod, or at the end point of the rod, where the rubbing against the paper occurs. This variation of the experiment provides a good means for separating what in the judgment of surface type depends on vibration experiences from what depends on resistance experiences.[3]

4. *Order of magnitude of differences in level on rough surfaces.* A few remarks will be made here on the order of magnitude of those differences in level that determine the roughness character of surfaces like the papers used here (Katz, 1923a, p. 2 f.). Since for technical reasons, the appropriate measurements could not be carried out in simple fashion on the papers, I constructed measurable tactual surfaces myself in the following way. Part of the surface of a glass plate of known weight was etched just so heavily with

hydrogen fluoride gas that the tactual impression of the exposed region of the glass could just be discriminated from the surrounding (smooth) part. The weight loss of the glass plate was determined as .000749 g on a microbalance, so that at a specific weight of glass of 2.5, the volume of dissolved glass amounted to .3 mm^3. The etched surface of the glass was approximately circular, corresponding to the opening of the platinum crucible that served for the mixing of the chemicals, and covered approximately 2800 mm^2, so that it lost a layer having a mean thickness of .000107 mm. Even should the hydrofluoric acid have attacked different portions of the glass in a very different way, it is safe to assume that the resulting differences in level were far less than .001 mm. This determination of limits is very crude, but it fully suffices for present purposes. Those differences in level that determined the roughness character of the smoother papers in our series and made their mutual discrimination possible, are similar in order of magnitude to the differences in level ascertained for the etched part of the glass. I might even venture to state that Papers 1 and 2 in our series each had smaller differences in level than the etched part of the glass surface. Paper 3 is probably about equivalent in roughness to that part. The more we approach the other end of the series of papers, the more the papers exceed the etched glass in roughness. From everything now known about the spatial sense of the skin,[4] we may well assert that the differences in level found in most of our papers would no longer be apprehended as such with a moving touch organ. Only Paper 14, the black cloth, provides a definite exception to this.

Practice presumably would improve the recognition of slight differences in level far beyond the ability of the normal subjects discussed here. What seems most astonishing in this connection, is the performance of Willetta Huggins, an American girl who became completely deaf and blind at 14 to 15 years of age. She could recognize the numbers on paper money and the large headlines of newspapers by mere touching.[5] Indeed, the refinement of the sense of touch, brought to a peak by practice, may explain many performances by mediums who claim to have a telepathic power.[6] In the future, this source of error must always be considered when telepathic recognition of written letters and the like is supposed to be attained only by touch, and precautionary measures must be taken accordingly.

The individual dots that form letters in the mechanically reproduced braille writing for the blind, which are 1.2 to 2.1 mm in diameter and .5 to 1 mm high (Buerklen, op. cit., p. 120), are huge bumps compared with the tiny elevations on the papers just mentioned. These dots indubitably feed into the *spatial sense* of the skin. To grasp them we produce quite reflexively a much different type of movement than when we explore for differences in roughness. Only slow movements, systematically varying in all directions, make such small forms available to the spatial sense. Roughness, on the other hand, is discriminated better with rapid movement, and systematic variation in the movement in all directions is no longer necessary. The properties of surface texture are most rapidly recognized at a pressure level that is too great for discerning spatial forms. An unpracticed person, though, does not do especially well, even if he has adjusted himself as well as possible to the stereognostic task. To cite a few examples, by no means can one readily discriminate the g formed with four points from the h formed with three points, or the p formed with four points from the q formed with five points. Given the breakdown of the stereognostic performance of the skin sense when confronted with that sort of crude difference in stimulation, how out of place would it then be to expect the spatial sense of the skin to be able to discriminate differences in level on our papers. That we must speak here of two completely different capabilities is indicated also by the fact that the touching movements that endeavor in vain to discern the sequence of braille letters, nevertheless spontaneously provide us with information on the roughness of the paper upon which they are pressing. Introspection shows us that in the normal subject the attempt to recognize letters of the braille alphabet (one can also do the test with a medal or with a monogram pressed into paper) involves reproducing visual images that then lead to recognition. In a sighted person, recognition is always via this detour, but the roughness of the papers is definitely ascertained not by visual mediation, but tactually. I should yet like to pose a question that certainly only the expert can resolve, which is whether in teaching the blind, the great sensitivity for differences in roughness could not be exploited more fully than hitherto has been the case.

Establishing the fact that differences in level pertaining to a microscopic world have an effect on the sense of touch is theoretically significant in several respects. First of all, this fact means that remote touching can no longer be regarded as occurring only in the experiments in which touching was carried out with a wooden rod. Even the experiments described above involving collodion and adhesive tape must be included here, since the leveling effect of the media definitely prevents the sensory organs lying underneath from obtaining even approximately corresponding pressure images from most paper surfaces. Thus, remote touching is also involved here. This suggests, however, that the basis for judgment is the same as in the experiments with the wooden rod: vibrations transmitted by the medium from the friction point to the sensory organs. It takes only a small further step in the same direction to suppose that the basis of judgment in touching with the bare finger is the same. The strong, horny layer [or stratum corneum] of epidermis at the end of the finger, whose tactual performance was discussed in the basic experiment, has a sufficient leveling effect so that the differences in level, at least in the smoother papers in our series, are no longer available to the sensory organs as pressure differences.[7] We will pursue the theory of tactual capabilities no further at this point. In concluding, it should simply be noted that the results of the following experiment also make sense from the viewpoint of remote touching, that is, by (and only by) the efficacy of the vibrations. We varied the basic experiment so that the papers no longer were touched in the customary manner with the volar side of the end member of the finger, but with the dorsal side, that is, with the nail. Even with such an experimental procedure, most differences existing between the papers were recognized.

5. *Fluid intermediaries.* In Section 9, the question was posed as to what varieties of tactual impressions occur in the case of wet, oily, sticky, etc. surfaces. We will now pursue this question with a few experiments. I put a bit of glue (Syndetikon) onto my finger and touched the papers of the basic experiment. All differences were obliterated; the smoothest paper could no longer be discriminated from even the roughest, the cloth paper. One has no idea of what sort of material could be involved, so there is a reduction of the second degree. The possibility that the glue quickly disintegrates on the surface of the paper can be dismissed, considering the brevity of the action—a single rapid touch. The thickness of the layer of glue is difficult to estimate, but it may safely be assumed to be insignificant at the moderately heavy pressure used in the touching, and it presumably was no thicker than the previously produced collodion layer. The bits of glue may have resided predominantly in the furrows between the epidermal ridges, scarcely preventing the ridges themselves from coming into direct contact here and there with the papers. But we know very well that diaphragms of quite different measure need not, in and of themselves, produce any impairment in tactual performance. How then can we explain the reduction that occurred in spite of such seemingly favorable experimental conditions? The answer is not difficult to find, and it is of considerable theoretical import. In touching with Syndetikon, no sounds occur; nothing can be heard even when the ear is brought as close as possible to the tactual surface. Thus, the glue acts as a lubricant that permits no vibrations to occur during rubbing, and consequently the sensory organs beneath the epidermis cannot be excited by vibrations. One can hardly imagine more striking proof that the vibrations which occur in touching are crucial for the judgment of roughness. As to the positive aspects of the impression of stickiness, I agree in all essential respects with the description that Zigler[8] recently provided: 1. Stickiness is a simple tactual pressure or contact sensation having certain characteristic features: the intensity of the contact or pressure increases gradually, then rather suddenly decreases to a minimum; the duration of the sensation is perceptibly longer than that of simple pressure. 2. Stickiness includes both the deeper-seated pull as well as the superficial light contacts that are set up when the sticky stimulus separates from the skin. 3. Stickiness is aroused only by a moving stimulus, although the movement need not constitute any

conscious feature of the sensation.⁹ This description concerns the sticky intermediary, but what about the tactual surface touched through it? The surface appears to be the same for all papers and to be very smooth. If one dilutes the glue somewhat, then the coarser differences in the paper can be discerned. The further one dilutes, the more the finer differences emerge as well, and the clearer the sounds of touching become. The papers are apprehended through the veiling stickiness in a way that is difficult to describe. Stickiness is a *state*, not a *property* of the paper. The state of the papers is perceived as that into which they have accidentally entered, just as an object can enter into light or shadow. The experiment with undiluted glue also shows that the force exerted in touching is not crucial for the impression of roughness, because the frictional resistance that must be overcome due to the viscosity of Syndetikon is considerable, but, as already noted, the tactual surfaces appear to be very smooth.

With a sticky medium it is easy to separate what capabilities can and cannot be attributed to the spatial sense of the skin in touching. If one uses so-called moiré papers, that is, paper that bears a raised pattern similar to that of silk moiré, then this pattern, whose elevation may amount to about .2 mm, can be recognized quite well in spite of the glue, scarcely worse than without the glue. Here, therefore, is one capability of the spatial sense of the skin that does not depend upon vibrations.

I will forego describing in detail the results obtained using other fluid media (water, oil, petroleum). Basically, they offer nothing new, but confirm time and again that the tactual impression is decisively determined by the influence the fluid medium exerts on the vibrations that occur in touching.

Footnotes

1. A few fingerprints are shown below (Section 35), which bring out quite clearly the design of the epidermal ridges. It is well known that the epidermal ridges are used in the identification of individuals in forensic anthropometry.
2. As to the extent of assimilation of differences, see D. Katz, Luftlicht und Beleuchtungseindruck (Air-light and the impression of illumination). *Zeitschr. f. Psychol.*, 95, 1924, p. 134 f.
3. H. Hoffmann, *Stereognostische Versuche* (Stereognostic experiments), dissertation, Strassburg, 1883, raises the question (p. 141) as to whether the contact sensation is necessary for the perception of the impressions that we term smooth and rough. The experiments above indicate a negative answer to this question, if the contact sensation is meant phenomenologically. In the same place, Hoffmann mentions that a female patient of Puschelt with no stereognostic ability recognized smoothness and softness in objects.
4. On this point, see, e.g., the relevant data in V. Henri, *Ueber die Raumwahrnehmung des Tastsinnes* (Spatial perception of the sense of touch). Berlin, 1898.

5. J. Thomas Williams, Extraordinary development of the tactual and olfactory senses compensatory for loss of sight and hearing. *Journal of American Medical Association*, 79, 1922. In the history of the psychology of the blind, reports recur in which the blind are able to give information on the color of cloths and the like which they touch. A consideration of all the facts leaves no doubt whatsoever concerning the self-deceptions of the blind in these cases; their feats consisted of associating the different individual tactual impressions with the colors that had been named for them by sighted acquaintances (see K. Buerklen, op. cit., p. 39 f.).

6. In the book, *Das Okkulte* (The occult), by Count Hermann Keyserling, Count Kuno Hardenberg, and Karl Happich, Darmstadt, 1923, there is a report on the capability of one medium, who is represented as being extraordinarily amazing. It is said (p. 93 f.): "In a state of trance, with fully or almost fully shut eyes, and the eyeballs turned upward, vision was quite impossible. I put into his hand a brochure whose outside bore the title, 'Ernst-Ludwig-Heilanstalt' in letters approximately 3 cm high. The letters were no different in level than their background, and one could not have touched and thereby read them by attentive movement of the fingers over them. I then put this brochure, which he was not familiar with and had never seen before, on his knee, put the fingertips of his right hand onto the title, and asked him to concentrate now on his fingertips so that he could read the writing with them. He first moved uncertainly over the letters, visibly pressed more and more on them, attempted mightily to concentrate, and became greatly agitated. After a period of the greatest effort, which increased more and more, he produced the first letter, E. Suddenly, in an ever increasing excitement, he snatched up the brochure, pressed it tightly against his stomach, and then began with almost convulsive effort and groaning to spell very rapidly, as if what he had apprehended could again escape from him. The letters were printed as uppercase Roman letters, the s in Ernst in the following form: S. He then said: e - e - e -, Emil, no, e - e - m -*question mark* - t - e - a period - N. t - E - r - n - *question mark* - an St. - Ernst; then he began to count very rapidly: 1, 2, 3, 4, 5, 6 letters - uppercase L - 1. 2. 3 uppercase and 3 lowercase letters - L - ... L... w - i - g - Ernst Ludwig. Because of the extraordinary convulsive effort that this experiment produced, I stopped at this point." The performance was interpreted as "stomach reading," as has already been observed by other mediums. I personally do not doubt that it involved reading with fingertips that had become very sensitive due to the trance. Willetta Huggins, who was mentioned above, did the same and more—even without being in a trance!

7. According to Drostoff, the epidermis at the fingertip has a thickness of .76 to .78 mm. F. Kiesow and M. von Frey, Ueber die Funktion der Tastkoerperchen (Function of the tactual cells). *Zeitschr. f. Psychol.*, 20, 1899, p. 162.

8. M. J. Zigler, An experimental study of the perception of stickiness. *American Journal of Psychology*, 34, 1923, pp. 73-84.

9. I note here three additional points in Zigler's analysis: 4. The different kinds of stickiness have their basis partly in visual associations, partly in the tactual present. 5. The perception of stickiness shows no essential alteration when the

touch organ is passively moved by another person. 6. The perception of stickiness disappears on a portion of the skin rendered anesthetic by ether.

Section 25. Remote Touching, Especially in Medical Practice

Lotze has masterfully depicted the extraordinary variety of remote touching in his *Microcosmus*. A few samples: "We fancy that we feel the contact of the rod with the object at a distance from us as directly through sense as we do its contact with the surface of our hand. . . . Only on this condition is the stick with which he gropes of use to the blind man or the probe to the physician. . . . In sewing we seem to be immediately percipient at the point of the needle."[1] Somewhat tongue-in-cheek, Lotze stated what significance our articles of clothing, which act according to this principle, have for expanding the space infused by our bodily self. In the age of the automobile and airplane, it is by the same principle of external projection that the automobile driver feels the goodness of the road via the tires, and the pilot feels the elasticity of the air via the wings of the airplane. When Lotze describes the aforementioned phenomena with the words, "all this we fancy we do not discover by reasoning—sensation itself seems to contain full knowledge on all these points (op. cit., p. 589)," then we see expressed here a phenomenological sensitivity, which was characteristic of Lotze, and which saved him from the natural temptation to rely on unconscious inference for the purpose of explanation. Remote touching with stereognostic intent, which looms so large in our everyday activities, and which renders services scarcely suspected by most people, achieves astonishing refinement in the physician, owing to systematic practice. The dead instrument comes alive in the disciplined hand of the physician, and provides information on the finest structures in body cavities, which are inaccessible to the hand on anatomical grounds or because of the danger of infection. "One learns very quickly . . . to touch just as well with the curette as with the fingers, and to feel whether portions of the placenta or of the fetus have been left behind."[2] With the war amputees, the orthopedic surgeon faced a completely new type of problem, which ought to be considered here. After all, touching with an artificial limb also occurs according to the probe principle. The so-called sensitive prosthesis especially aids stereognostic activity (Katz, 1921, p. 5 f.). The internist's palpation involves a different method, and also has a different psychological structure. All textbooks on methods of examination[3] regard palpation to be an indispensable diagnostic aid for the physician, but on closer inspection one finds surprising differences between the methods in their qualities (*Modifikationen*) and their apparent effectiveness.

Hausmann appears to have brought palpation to a height of development scarcely yet surpassed.[4] The reliability of his diagnoses must be credited to his tenacious, clever, and exhaustive practice with all possible combinations of types of touch (the physician must "palpate thinkingly and think palpatingly"[5]). Considering what we have established about the microlevel capabilities of the sense of touch, even at a distance, the general doubts occasionally expressed about Hausmann's methods are misplaced. When he himself calls palpation the most subjective of all methods of examination, the meaning of the subjective training factor is thereby correctly captured. To achieve good results, according to the practitioners and particularly Hausmann, one must carefully follow the directions for executing the tactual movements during palpation. There are instructions, for example, on the extent and direction of movement ("the gliding of the fingertips past the organs, or the organ past the fingertips, considerably heightens, during the moment of movement, the precision of the tactual sensation") as well as the natural teamwork of the fingers (Hausmann mentions Pcllatschek's octodigital palpation). These instructions show that much the same holds for achieving the stereognostic performances as for the tactual performances discussed hitherto. Much of what the experienced practitioner produces with palpation, using the most appropriate finger positioning, could well lead to new psychological research.

Goldscheider and Hoefer also dealt with the psychology of palpation in their research mentioned above. "What usually is involved is the touching of deeper-lying, tougher tissue through layers of impressionable tissue. During the requisite compression of these layers, the touching skin itself is impressed. Thus, in penetrating into the deeper layers of tissue the examiner receives steadily increasing pressure sensations, and then a greater increase when a more resistant, deeper-lying structure is encountered."

In feeling the pulse, such properties as frequency, rhythm, and speed of pulse may be judged, along with the state of the artery wall. These properties presumably are revealed not only, or even primarily, to the spatial sense of the hand, but also require the participation of vibration sensations. And vibration sensations also may play a role in so-called palpatory percussion. "In utilizing the percussion effects, one is quite instinctively influenced by the resistance felt by the finger carrying out the percussion, as well as by the sound perceived" (Sahli).

Yet to be discussed (Division III, Chapter I) is the affinity of sound and vibration sensations, which leads to confusions between the two sensory modalities under some conditions, but which may help clarify the essence of palpatory percussion, on whose *sensory basis* not even the outline of an investigation has been completed. "Be that as it may, there is yet something instinctual about this mixing of sensations, even something unconscious, which presents great difficulties for its scientific analysis and its use in a didactic relationship, since it is not easy to teach a learner how to direct the attention simultaneously during percussion to two sensations that are difficult to judge (Sahli)." In addition to the spatial impressions of the skin and vibration sensations, the masseur presumably also relies at the same time on the perceptions that only arise from the effort he or she expends in the muscle layers, sinews, and tendons.[6]

Footnotes

1. Lotze, op. cit., p. 588.
2. Schottmueller, Das Problem der Behandlung infizierter Aborte (The problem of treating infected abortions). *Muench. med. Wochenschr.*, 1921, p. 663. According to the author, the mortality rate decreased by about 90% when the curette was used in place of the hand.
3. Sahli, *Lehrbuch der klinischen Untersuchungsmethoden* (Textbook of clinical examination methods). Sixth ed., Vol. 1, Leipzig and Vienna, 1913.
4. T. Hausmann, 1. Die methodische Intestinalpalpation (Methodical intestinal palpation). Berlin, 1910. 2. Die methodische Gastrointestinalpalpation (Methodical gastrointestinal palpation). Berlin, 1918. 3. Beruehrungsempfindung und Druckempfindung, insbesondere die tiefe Druckempfindung (Contact sensation and pressure sensation, particularly the deep pressure sensation), *Pfluegers Archiv, 194*, 1922.
5. Boas, *Diagnostik und Therapie der Magenkrankheiten* (Diagnosis and treatment of abdominal illness). Leipzig, 1911, p. 33.
6. A. Hoffa, *Technik der Massage* (How to massage). Stuttgart, 1893.
   Editor: Katz (1936) reported that Hausmann performed palpation with the fingers anesthetized and still obtained entirely satisfactory results. Thus, the recognition of shapes in palpation "is at least partially controlled by the sense organs in the muscles, the sinews and the joints" (p. 148).

Section 26. Tactual Performance of Body Parts Not Usually Used for Touching

1. *Touching with the lips.* When faced with some touching task, then as sighted people we almost invariably use the hand, or more precisely, the fingers. Most of the remaining body surface is typically covered with clothing, and therefore cannot be brought into contact with other tactual surfaces. As for the head, although it is free of clothing, we almost never use its parts for touching. The inside of the mouth, mainly the tongue, comes into contact with all foods, and the lips with many foods, but who would use his or her own tongue and lips to select papers, types of wood, leather, etc.?[1] It is conceivable that it is done, but that is not to say that anyone has ever done it.

In an initial set of tests, the red portion of the lips was used for touching. The experimenter bent papers together into an approximately cylindrical shape, and moved them to and fro with the curve on the lips, so that neither the face nor any other part of the subject's body was touched. The subjects were not allowed to handle the papers, because they would have recognized them with their fingers. What were the results of the experiment? Subject L not only distinguished all papers from one another, as in the basic experiment, but in addition detected yet other differences in the papers that had eluded her fingers. For example, in the comparison of 9 and 10, not only was 10 designated as rougher, but it also was said to have many more little hairs; whereas 9 was called roughened blotting paper, 10 was called patterned drawing paper. When asked to compare this touching with that in the basic experiment, L indicated that more details were felt by the lips. She said that she undoubtedly had occasionally brought objects to the lips for touching, but that she had never practiced this systematically. The other observers in this experiment, my wife (Ka) and D, also correctly discriminated all papers of the set from one another. As in all experiments, I checked this one on myself and stumbled onto a curious source of error in so doing.[2] I found that almost all of the papers smelled different, though these differences in odor probably did not aid the judgments on the tactual impressions.

2. *Touching with the toes.* In a second set of tests, the large toe of the foot served as the touch organ. As in the basic experiment, two surfaces were placed before Subject L, who actively touched them one after the other with her toe. The result was very surprising. Ten correct judgments were obtained, no difference was discerned in two cases (1 equalled 2, and 5 equalled 6), and one case can be called incorrect (12 was called smoother than 11). Similar results were obtained with Ka and myself. The large number of correct cases shows that the toes fall little behind the fingers in their tactual capability. This comes out particularly in the fact that L did not limit herself to relative judgments, but also provided a number of absolute judgments, as, for example, in characterizing 4 as a non-glossy, very fine cardboard, 8 as granular, 13 as a strong drawing paper, and 9 and 10 as fine, soft blotting papers. The foot, which normally walks clothed in wool, cotton, and occasionally silk, no doubt sometimes comes into bare contact with wood, carpeting, linen, sand, and stones, but rarely with paper and most certainly never with most of the papers that it now discerned correctly. Our experiments show that the capability of a touch organ seems to depend little upon its particular prior experience. This holds first of all in the ability to discriminate roughness. Does this finding not suggest that the basis in sensation for the judgments made here is to a large extent independent of experience and has a somewhat elementary quality, at least in the sense that we think of the pressure or two-point threshold of some part of the body as independent of experience? These experiments provide an important hint

in interpreting all of the tactual performances treated here. When papers touched by the toe are characterized as drawing paper, blotting paper, etc., then *that* is naturally inconceivable without some kind of prior experience. What is remarkable is that the experience gained in touching with the fingers and expressed in their characterizations, now also serves to interpret the tactual experience on the foot, where the corresponding prior experience had not been obtained. This indicates that the *absolute* impression of the papers on the fingers is closely related to their absolute impression on the toes, thus enabling a reproduction of the corresponding designations.[3] Naturally, the results just discussed do not contradict the previously stated supposition that the tactual imagination, which presumably produces the image of a tactual surface on a portion of the body that never had the corresponding tactual impression, works with a memory touch. Further experiments using the toes will follow below.

3. *Touching with a rod in the teeth.* In order to radically curb the influence of special types of experience, I also carried out experiments in which the touching was with a rod in the *teeth*. Who would ever have resorted to the teeth in touching papers? Touching was carried out with a wooden rod, about 12 cm long, pointed at one end, and held with the teeth. To make the rod easier to hold, its other end was provided with a material that dentists use to make impressions of teeth, which the subject bit into. This prevented movement of the rod during touching. The 14 papers were presented just as in the basic experiment, and the subject touched with the eyes closed. The arrangement did not force the head into an uncomfortable position. In a blind test, I myself made ten correct judgments, and could not discern any difference three times. My wife produced the same number of errors, but they were at other positions in the series. Similar results were obtained with other observers as well. Many judgments were based on differences in the resistance experienced while moving the head. Most often by far, however, particularly with small differences, judgments were based on differences in the vibration sensations, which were localized in the teeth. As in the experiments cited above (Section 24, Subsection 3), the vibrations were produced by rubbing the wooden tip on the papers, and they were transmitted to the teeth through the rod. The results of the experiment confirmed the irrelevance of prior experience in the discrimination of vibration sensations, on which the judgment of roughness is based. As in the experiments with the wooden rod described above, the material here also was not recognized in an unambiguous fashion. One could imagine it to be paper, but it could also be other material.

It must be revealed that there was a source of error in the present experiment that could not be eliminated. The sounds produced by the friction between the wooden rod and the tactual surface were transmitted to the ears by bone conduction and were perceived as such. They could not be eliminated; stopping up the ears naturally does not help, but, on the contrary, makes the noises even more noticeable. The noises were entirely unique on many papers, but differed little on others, and one could undertake to classify the papers based on the noises they produced. The task posed for the subject, however, was to direct his or her attention to the vibration sensations themselves and use their differences as a basis for the judgment on roughness. To be sure, it must be left open for now as to whether the character of the sound did not occasionally influence the judgment on roughness. I have this source of error to thank for a theoretically valuable finding. When the fingers hold the rod and use it to touch the smoothest paper, then the noises thereby produced are so weak as to be barely audible to an ear placed close to them. If, however, the rod is held with the teeth, then, because of bone conduction, those noises are much louder and their distinctive features stand out much more prominently. This source of error thus provides a means to diagnose vibrations, which will find use in supporting an important portion of our theory of tactual performance.

Footnotes

1. To be sure, housewives sometimes use their lips to test fabrics.
2. This source of error also may help to explain the so-called telepathic powers of mediums.
3. Neither in these experiments with papers nor in those which follow below (Sections 29 and 30), which were carried out with other, very different materials, did any results whatsoever indicate that there is an analog on micromorphic [substance] tactual forms [e.g., roughness, hardness] to the so-called Praegnanz (precision) of visual gestalts in Gestalt theory. It would be of great importance for Gestalt theory to establish whether the macromorphic [shape] tactual forms betray a tendency for Praegnanz.

Section 27. One's Own Body as a Tactual Object

A few thoughts are inserted here on the case where, instead of an alien object, a part of one's own body is made the object of touching. The mutual perception of two body parts having the same sensory organs is unique; there is nothing like it outside of the skin sense. E. H. Weber suspected quite rightly that these experiences provide an important impetus for the vitalization of inanimate nature with causally effective powers. "In what other sensory organ do we have a similar ability . . . to become conscious of the causal connection than here, where we become conscious of the effort of will when we ourselves press one hand against the other hand (op. cit., p. 93)." We will highlight just a few out of the lively multiplicity of these tactual impressions. First of all, we are reminded of Weber's finding cited above, that the moving part of the body feels the motionless part as object; in our terminology, the objective pole prevails. The subjective pole completely predominates in the impression received by the motionless touch organ. Which particular pole is emphasized in each of the two tactual experiences is independent of which experience stands out as a whole. In the dual-touch situation, with a completely neutral attitude, the tactual experience involving the objective pole may stand out, while that involving the subjective pole furnishes the background, but this relationship can be reversed with a different selection of positions touched.

We will limit ourselves to the case that occurs most often in everyday life, in which the touch organ is the moving hand, and the portion of the body it touches varies. In addition, a neutral, unconstrained attitude of attention is presumed in the following. If I touch my finger to the edge of my tongue, which is held tightly between my teeth, then the tongue appears as object, even though the experience of contact on the tongue predominates completely, and, indeed, with gentle touching we must make an effort to discern the tactual experience of the finger in the overall phenomenon. If you run your fingers through your hair, then the tactual object "hair" absolutely predominates in the dual-touch experience, while the diffuse touching experience on the head stays completely in the background. I have selected extreme cases here, since touching the hand to other parts of the body produces touch complexes in which both of the component impressions assume an intermediate position relative to the two directions indicated. The influence of attention is remarkably strong; it can shift considerably the relative salience of the two component impressions. One can also make *both* tactual experiences clearer through attention, and allow them to stand side by side with equal emphasis, just as can be done with the tones in a dual-tone sound. This grouping succeeds best, assuming some practice, in cases where the two tactual experiences show the greatest contrast in their polar structure. I have found that the greatest difficulties occur in analyzing those double impressions that are obtained when the fingers of one hand touch the same fingers on the other hand, or when the fingers of one hand oppose each other and thus mutually touch. It really is not very easy to say which finger is tactual object for what other finger in such a case, the unity of the touch complex being carried to an extreme here. This touch complex is of some significance for certain capabilities of the touching hand discussed below (Section 31). If one finger in the last-cited case had been covered with collodion or adhesive tape, then it would become the tactual object in a pronounced way.

Section 28. Experiments with Adapted Touch Organs

1. *Adaptation by rubbing.* A sensory organ can be fatigued by prolonged, heavy use; one way the fatigue reveals itself, among others, is by a change in

the reaction to stimuli. The senses fatigue at varying rates: smell very rapidly, the ear most slowly and least of all, although even its fatigability can be definitely demonstrated. We touched on the question of the fatigability of the sense of touch earlier [Section 15], when we demonstrated the necessity of distinguishing between fatigue for moving and motionless stimuli. The concept of adaptation takes precedence over that of fatigue, because there are cases of adaptation that cannot be attributed to fatigue. Prolonged wearing of a glove produces adaptation, but not fatigue, as already mentioned. Some rather unsystematic experiments on the adaptation of the touch organs will be discussed here.

Rubbing was one means that I used to adapt the fingers; here, of course, fatigue was involved. So-called corrugated cardboard, for example, proved very suitable for the task, which was to energetically rub the finger to be adapted to and fro. Depending upon the duration of the rubbing and the pressure exerted, the finger can be fatigued to a varying extent. The rubbing can be pushed to such an extent that one can discern neither differences between materials nor their qualities (*Modifikationen*), and thus the fatigue has produced a reduction of the second degree. With the lightest degree of rubbing, all touched surfaces feel different to the finger, but discriminability is scarcely affected. Precise quantitative tests cannot readily be carried out, because the extent of fatigue after heavy rubbing changes within a short period. Rapid recovery from the rubbing is probably due to the concomitant artificial hyperemia [congestion of blood] in the finger.

In an initial experiment, I rubbed the tip of my index finger over corrugated cardboard for about two minutes, but then had to stop because the fingertip had become painfully hot. If I then touched an object with my finger, I felt nothing at all at the point of contact; the fingertip had lost its sensitivity. If I pressed more heavily, then I noticed only the strain in the joints of the finger. If I rest the finger, then I can feel how the sensitivity recovers, the stiffness decreases, and the touch sensation returns. I again deaden the finger by rubbing and then move it over very different tactual surfaces. It is like moving a finger wrapped in a thick woollen material over the surfaces; not the slightest difference is perceived. However, if I prolong the touching over the same surface, so that the finger recovers, then the tactual impression becomes ever fuller and more distinct. I also carried out a few such tests with Subject L. After about 30 seconds of rubbing, she was supposed to compare two papers at a time with each other, as in the basic experiment, and do so rapidly, so that the adaptation changed as little as possible. Papers 1 to 6 could no longer be discriminated from one another, Paper 1 was discriminated from Papers 7 to 14, Papers 7 to 13 appeared about the same, but all were discriminated from Paper 14. These experiments should suffice to illustrate the procedure with rubbing.

2. *Adaptation by cooling.* Just as with rubbing, any desired level of reduction can be achieved by cooling the touch organ. One cannot speak of fatigue in this case. The subject must hold his or her hand in snow until a very intense, thereupon highly painful, cooling has occurred. Before making the touching movement, the fingers are quickly dried with a cloth. In these experiments, each paper in the set was judged by itself. After each judgment, the hand was thrust back into the snow in order to return the rapidly disappearing cooling to its previous level. After very intense cooling, Subject L reported for most papers: "I slide over something that does not offer too much resistance, but cannot say what it is." All differences between the papers have disappeared, and even the material itself is no longer recognized. Somewhat more specific statements were possible with lesser levels of cooling, and I report them here: 1, 2, and 3. Very smooth. 4. Smooth. 5. Smooth, not for writing. 6. Rougher than 5. 7. Smooth surface. 8. Somewhat rougher surface. 9. Not as smooth as 7. 10 and 11. Smooth drawing paper. 12. Impression of fibrous paper. 13. Like 12. 14. Rough, small fibers. The extent of reduction can be gauged by a comparison with the statements in the basic experiment. There was a quite pronounced tendency for the subject to judge the papers as smoother, which contradicts the hypothesis that the impression of smoothness depends entirely upon how much force is exerted during active movement. (It seems hardly plausible that the frictional coefficient between the skin and the tactual surface decreases as the skin is cooled; rather, the opposite would hold as a result of the incomplete removal of the moisture from the snow water.) This tendency also was evident for Subject K, whose statements were as follows in the case of an intermediate level of cooling: 1 and 2. Very smooth. 3. Glossy paper. 4. Very smooth. 5. Medium rough, somewhat woody. 6. Medium rough. Cloth? 7. Extraordinarily smooth, writing paper. 8. Moderately smooth. 9 and 10. Rough, cloth-like. 11. Rough, cloth paper? 12. Very rough, cloth-like. 13. Very rough, quite coarse cardboard. 14. Very coarse and woody. Here, too, the extent of the reduction produced by cooling can be seen by comparing these judgments with those obtained in the basic experiment. The illusion of greater smoothness can very probably be explained by the fact that cooling reduces the sensitivity of the sensory organs to vibrations.[1,2]

Footnotes

1. R. Allers and F. Halpern found that the touch threshold was lowest at a skin temperature of 36 to 38 deg C, and that further warming of the skin increased the touch threshold. Die Beeinflussung der Tastschwelle durch die Hauttemperatur (Influencing the tactual threshold through the skin temperature). *Pfluegers Archiv, 193,* 1922.
2. Editor: S. Lederman (personal communication, June 1988) provided an alternative interpretation. If the skin is cold, it cannot deform as much (in keeping with the deformation theory of Taylor & Lederman, 1975).

PART II: Experiments on the Identifying Characteristics (*spezifikationen*) of Surface Touch

Section 29. Recognition Time for the Identifying Characteristics (*spezifikationen*) of Surface Touch

1. *The tactual material.* Whereas papers were used exclusively in the preceding experiments, we will deal with a large variety of very different tactual materials in this part. First, we list all those materials used regularly in the experiments:

1. patterned, coarse linen	21. terrycloth
2. fine, unstarched linen	22. plush
3. starched linen	23. cotton
4. stiff linen	24. crepe de Chine
5. velvet	25. satin ribbon
6. corduroy	26. felt
7. coarse sandpaper	27. voile, mounted
8. fine sandpaper	28. oilcloth
9. patterned paper	29. rabbit fur
10. blotting paper	30. hard rough leather
11. writing paper	31. smooth leather
12. tin foil glued on cardboard	32. Russian leather
13. smooth aluminum sheet metal	33. kid leather
14. rough iron sheet metal	34. calico
15. glass	35. wood cardboard
16. moderately rough oak wood	36. rep
17. rough fir wood	37. coarse packing paper
18. smooth fir wood	38. corrugated cardboard
19. cheviot	39. flannel
20. woollen cloth	40. silk

A more detailed description of the properties of these tactual materials would take us too far afield, and also is not absolutely necessary for an understanding of the experiments. In none of the experiments described below were all of the tactual materials used. Nearly all of the tactual surfaces were 15 by 15 cm in size. Even where it is not explicitly noted, materials such as cheviot, cloth, terrycloth, plush, and cotton, which are easily moved aside, were mounted on a solid support of thick cardboard.

2. *Experimental arrangement.* To measure the recognition time for the materials, as well as to clarify the recognition process itself, the following experimental arrangement was devised. Using a simple balance, a so-called

platform balance, which is often used in the home, I had a technician attach a terminal that could be adjusted so that the slightest movement of the balance plates from a particular resting position would break an electrical circuit passing through them. One balance plate (a) held the tactual surface, and the other (b) was so weighted as to slightly overcompensate for the weight of the tactual surface and to make a good electrical contact at the resting position attained. Since an imbalance of 2 g on one side produced a definite movement, we generally made the excess weight on (b) only 30 g in the experiments, that is, a weight that could be overcome by a slight pressure of the touch organ on the tactual surface lying on (a). The balance was tipped from its resting position, and the electrical contact broken, at the moment when a weak pressure just exceeding that of the 30-g counterweight was exerted by the subject on the tactual surface. To keep the hand in contact with the tactual surface during touching, and not let the tactual surface get away from the hand, an adjusting screw was set so that balance plate (a) could move down only about 2 mm from its resting position. The subject's forearm rested comfortably on a cloth-covered support; the hand was held over the touch surface before the touching movement and was lowered upon command. The tactual surfaces were out of view of the subject, and the ears were again stopped. The electrical circuit through the terminal on the balance did not pass directly through the Hipp chronoscope, but rather through a Cattell relay, from which the arms of the clock were engaged in the regular manner. The clock ran continuously, starting at the moment when the electrical circuit was first broken. This could not be done without introducing a [lock-up] relay, because otherwise the circuit would have been momentarily closed repeatedly at the terminal, owing to the somewhat springy action of the balance. The clock ran only until the subject closed the circuit again by speaking into a voice key. Thus, the time elapsing between the contact with the tactual surface and the reaction of the subject was measured.

Three types of reaction time were measured. Experiment 1. As soon as he or she touched and depressed the tactual surface with the right index finger (the two occurred virtually simultaneously), the subject was supposed to react, regardless of whether or not anything could yet be said about the properties of the tactual surface. We thus obtained the simple tactual reaction time. Experiment 2. Touching again was with the right index finger, but the subject was to react only after having recognized the tactual surface from its particulars. The type of tactual movement was left entirely up to the subject, who was told that his or her statements would be more valuable, the more specific they were. Thus, the subject was to strive to say as much as possible, and to be satisfied with a general statement only if totally unable to arrive at a more specific one. Since the subject naturally did not want to have too long a recognition time, a compromise had to be made each time between the desire to react as quickly as possible and the desire to react as informatively as possible. Experiment 3. Touching was with all fingers of the right hand; otherwise the instructions were identical in Experiments 2 and 3, which both measured tactual recognition time. Comparison between Experiments 2 and 3 provided certain information on the difference in performance between one finger and the entire hand.

The experiments were carried out with 27 tactual surfaces. Since, as was to be expected, the simple tactual reaction times (Experiment 1) were about equally long for the various tactual surfaces, all surfaces were presented only once, and a mean was calculated from the individual reaction times. In Experiments 2 and 3, all surfaces were presented twice each. In order to give about equal practice, 9 tests of one type alternated with 9 of the other type. An absolute judgment was made of the individual tactual surfaces, not a comparison between every two tactual surfaces, as in the basic experiment with the papers.

3. *Preamble to the interpretation of the results.* In embarking on the experiments, it very quickly became clear that I could not provide a satisfactory interpretation of the results without first serving myself as a subject. Accordingly, I had an assistant present the 27 tactual surfaces in random order to me, with my eyes closed and ears stopped. Time was not measured; I was only interested in fathoming the process of recognition of specific surface touches. Based on my observations, one can distinguish the following types of recognition processes: 1. A judgment on the tactual surface is possible immediately, with no awareness of any separate sensation-like or image-like intermediate elements. Thus, just as one generally sees at first glance whether a material is wood, paper, glass, etc., one also believes he or she is directly touching the wood, paper, glass, etc. Once a confident judgment has been reached, additional tactual movements serve only to confirm the judgment. 2. One immediately has a confident judgment on whether the surface is hard or soft, rough or smooth, elastic or brittle, but the material remains unrecognized at first. When the touching is continued further, driven by the desire to overtake the material, the recognition can then effectively be achieved, and in a way similar to that in 1, with no definite intermediate elements determining the judgment. It would not yet be warranted to speak of an inferential process that leads to recognition. This cannot happen until Case 3. Suppose starched linen is presented. I touch it and say to myself: it feels like wood, but the texture is really too regular for wood. It is probably a stiff woven fabric, perhaps linen. In such cases, intellectual processes indubitably play a decisive role, with my knowledge of the properties of tactual surfaces indeed being utilized in more or less complex inferential processes.

I correctly recognized all but one of the 27 tactual surfaces, which naturally was due to the fact that I had already handled all of them, and thus in many cases my performance merely involved recognizing them again. A very noteworthy fact established in the tapping experiments (Subject Hornthal) is that accuracy increases dramatically when the set of tactual surfaces is briefly shown as a whole (the 27 tactual surfaces for a few seconds), before the touching is carried out blind. As sighted persons, we rarely make direct associations between linguistic labels and tactual impressions, so the label is not easily evoked by the tactual impression. The difficulty is considerably eased when the subject is allowed a glimpse to orient himself or herself as to just what tactual surfaces can occur. The linguistic labels are thereby primed and then, presumably through the mediation of the visual images of the surfaces, the labels are more readily evoked by the individual tactual impressions. Even if the other subjects approached the tactual surfaces differently than I, who had constantly dealt with them, it nevertheless seems certain from their statements that they, too, employed the three different recognition processes distinguished above.

4. *Quantitative experimental results.* Mr. Horney (H), a philosophy student, Miss Bockmann (B), a philosophy student, and Miss Schoer (S) served as observers. Summarized in the following table are the mean reaction times in seconds. (The relay time has been subtracted.)

Subject	1. Simple reaction	2. Recognition (1 finger)	3. Recognition (5 fingers)
H	.215	5.090	4.788
B	.315	5.982	5.679
S	.371	3.859	3.143

I will not discuss the individual differences. The means are based on both the time for correct (or equivalent) recognition, and the time that elapsed until the subject gave up trying to formulate an appropriate characterization.[1] What is meant by correct and "equivalent" characterizations of the tactual surface will be clarified below (Subsection 5). There were extraordinarily large differences in recognition time between the tactual surfaces. Thus, there was a 20-fold range on Case 3 for Subject H, the recognition time being lowest for glass (.656 sec), and highest for starched linen (11.990 sec). Naturally, it must give one pause to include such different values in a mean, but, on the other hand, we could not imagine reporting all of the numerical results *in extenso.* The individual values of the simple reaction time showed much greater agreement, as was to be expected. The values listed in the table correspond approximately to simple tactual reaction times cited in the literature. Deviations can be attributed simply to the different arrangement used here, i.e., the finger sought out the tactual stimulus, rather than began in motionless contact with the tactual stimulus. The difference between the first and second columns shows the additional time spent when the endeavor to recognize the tactual surfaces replaced that of simple reaction. Before we describe the stages of achievement in that endeavor, let us briefly compare the second and third columns.

The recognition times reveal a clear advantage for touching with five fingers, rather than one finger, the difference in time amounting to .302 (H), .303 (B), and .716 sec (S). The superiority of the five-finger formation emerges even more strikingly in a 25% increase in the number of correct reports. As was previously pointed out, the five fingers of the hand ought to be regarded as a unitary touch organ, and subsequent experiments will show the "how" of this collaboration in even greater detail. "Usually we touch larger objects with five or ten fingertips simultaneously, when we want to find our way about in the dark. We then receive five to ten times as much information at a time as with one finger."[2] These words by Helmholtz, which hold not only for touching in the dark, express the view that we should even regard the fingers of both hands as one touch unit. In my experience, that is quite correct when we are confronted with a stereognostic task, and it is particularly true for blind persons. If only one surface is to be touched, then we generally are content using the five-finger assembly of a single hand. Implicit in the formulation by Helmholtz is the supposition that in working with five fingers there is a sort of arithmetic summation of the performances of the five individual fingers. But that is really not the case. The experience that emerges from the central processing of the sum of individual impressions in many ways also signifies something qualitatively new, as compared with the impression of the individual fingers. The increase in correct judgments supports this interpretation.

5. *Verification stages in tactual perception.* When should a judgment by a subject on a tactual surface be called correct or appropriate, and when false or incorrect? Since it particularly involves the identifying characteristics (*Spezifikationen*), we must go into this question more extensively here, but without getting bogged down in it, having already touched upon the question in the basic experiment and its variations. In my book on amputees (Katz, 1921), a start was already made toward solving the question, to which practical psychology had directed us: "It definitely would be too strict to regard as incorrect every tactual judgment that is objectively false, for example, the judgment of wood as linen, which could scarcely occur in visual perception. There are surfaces that are nearly or completely equivalent tactually, but can be discriminated very easily visually. Thus, we must take as our reference the immanent observation in deciding the correctness and incorrectness of judgments of tactual surfaces (p. 69)." Two tactual surfaces will be called tactually equivalent, and the judgment that both are of the same material will be accepted as correct, if they can no longer be discriminated under the most favorable subjective conditions, i.e., given the most adroit tactual movements and the best conceivable disposition of the

attention. In that case, the limits of the capability of an immanent observation have been reached. The tactual judgment can then be corrected only with the aid of other perceptions caused by the surfaces, whose differences are allowed to emerge. But note well: the tactual phenomena themselves are not affected by this; the observation that sets in is transeunt [vs. immanent], if we may use this expression for brevity.

Given such transeunt observations, let us now distinguish the various stages in the verification of the judgment on tactual phenomena. Wherever we doubt the reliability of touch, as sighted persons we will call first upon vision. In most cases, surfaces that appear equivalent to touch can, under very favorable visual conditions, still be discriminated visually.[3] The eye does not readily confuse linen with wood. If poor illumination prevents visual checking, then I can use other means of checking. I try to press my fingernail into the material in order to determine its hardness, or to test its elasticity with my muscular strength. The new perceptions can lead to a verification or disavowal of the previous judgments. In addition, the noises produced by tapping on the materials can aid in their discrimination when other means fail. (For more on this, see the Addendum of Section 45). I have also sometimes succeeded in discriminating by odor; where the hand fails, many tactual surfaces may be distinguished with the nose held sufficiently close. If all of the sensory psychological measures do not lead me to the goal, then I must resort to physical science and chemical methods. The latter are applied at a different level of the theory of recognition; in place of the perceptual apprehension of qualities, is the apprehension achieved by inductive inference. How far such induction can be used depends on progress in physical-chemical research.

As a rule, we rely more on vision than touch, and thus call upon vision first for verification when doubts arise. The opposite course is often taken when the illumination is poor, such as at dusk. Imitations of precious materials (i.e., when marble, stone, and metal are replaced by wood, stucco, and pulp paper) may be exposed with the help of touch. In purchasing fabrics, too, when more than just the color is involved, one will always revert from vision to touch, in order to be on the safe side.

6. *Qualitative experimental results.* When evaluating the tactual performances obtained, one must never lose sight of the fact that qualitative judgments were involved. If the tactual surfaces were presented by pairs, then almost every surface could be discriminated from every other one. If one fails to consider how the present procedure differs from that of the basic experiment, then one can easily underestimate the performances now to be discussed. Some tactual surfaces were recognized by all subjects when touched with either one or five fingers; these included velvet, fine sandpaper, glass, fur, felt, and cheviot. Coarse sandpaper, oilcloth, rough wood, writing paper, smooth (aluminum) sheet metal, and cloth were recognized in al-

most all cases by the three subjects. The reliability of recognition cannot be due entirely or even primarily to the frequency with which the materials listed had been touched previously, because, for example, fur, sandpaper, rough wood, and felt are not materials we normally come into contact with. In this case, it is probably the novelty and forcefulness of the impression that imprints it on our memory and leads to recognition. Some forms are so unique that they actually need to be encountered only once in order never to be forgotten again. Cloth, cheviot, and writing paper may owe their advantage to the regularity and frequency of their occurrence. It is different again with glass, sheet metal, and oilcloth; they enter the list of the best-recognized materials owing not only to the specific form of their surfaces, but also to the obtrusive coldness that colors the tactual impression. The fact that the temperature impression can do this is clearly shown by the experiments we will report in Chapter IV (Sections 36 to 38). Glass and sheet metal, just like oilcloth, are no longer reliably recognized by touch when heated a certain amount above the temperature prevailing in their surroundings or that usually attributed to them.

The experiments uncovered a whole series of tactual equivalences. I mention first those that recurred rather regularly for all three subjects. Tin foil was identified as oilcloth, starched linen as cardboard or paper, voile as thin woollen material, and patterned coarse linen as cotton material. Besides these relatively few cases of tactual equivalence that recurred for all subjects, there were a very large number of confusions that changed from person to person, or even within the same person from experiment to experiment. I can give only a few examples of these confusions. Russian leather was mistaken for cloth, rubber, and cardboard. Satin ribbon could be called calico, tissue paper, smooth leather, cotton material, and shirting. Blotting paper occasionally appeared tactually equivalent to leather, wood, and cardboard. And indirect confirmation of the importance of temperature impressions in the recognition of materials, cited above, is provided by this last series of confusions: the rough iron sheet metal was occasionally regarded as plaster, slate, and bumpy glass, that is, as a material having approximately the same temperature impression. As the experimenter, I was greatly astonished by many confusions, such as, for example, when satin ribbon was called smooth leather, or when Russian leather was called cloth. The astonishment seems to be based on my knowledge of the total dissimilarity of the visual impression and the chemical-physical nature of the confused materials. However, the confusions become more understandable when you take the position of the subject and have an assistant present the tactual surfaces. You then confirm that you could be susceptible to quite similar confusions. It is an important fact that the dissimilarity of the visual impressions can make the tactual impressions appear different to some extent, even when the hand alone notices no differences, and one thus must speak of the strong effects of visual associations. I must again ask the reader here as well to illustrate what has been said by a few tests of his or her own.

The instructions did not permit the subjects to use the means enumerated above to make transeunt verifications of their judgments and thereby to recognize tactual equivalences for what they are. The type and extent of tactual movement was left up to the subjects, but they were forbidden to use their fingernails as scrapers in order to obtain information on the hardness and softness of the material. The subjects could only endeavor through increased concentration of their attention and the most appropriate movement possible of the fingers, to bring forth a more detailed and comprehensive statement on the tactual surfaces. And observation of the fingers, particularly in the difficult cases where the recognition time rose to 10 sec and more, shows how the subjects attempted to use their one opportunity to improve the judgment, sometimes by moving nervously to and fro, sometimes by slow, forceful rubbing.

In the recognition of a material such as cloth, glass, oilcloth, leather, etc., does the tactual impression itself trigger the process of recognition that eventually yields a linguistic label, or does it merely evoke the corresponding visual image, which in turn determines the process of recognition? I never questioned the subjects about this, nor do their spontaneous statements provide a general answer, but this much is certain: in many cases, very clear visual images occurred, and it is probable that these participated in the recognition process. It is even conceivable that here and there strong visual images somehow arose that put the subjects onto the wrong track, and made subsequent correction by the tactual impression impossible. For visual images that are tied to specific personal experience, there are spoken expressions such as "colored paper," "glossy leather," "probably black oilcloth," and "the material is presumably not white." How lively the visual images can be was shown to me by tactual experiments with untrained amputees (the report on this is given below in Section 30), where, with the eyes shut, the categorical judgment was made that the cloth just touched was red. These subjects had not the slightest awareness of the comical aspect of such an assertion.

7. *Experiments with variation in intensity of pressure.* While the platform scales permitted the subject to exert any preferred level of pressure on the tactual surface, the following arrangement was designed to control the level of pressure on the tactual surface in order to determine the significance of the pressure variable for tactual performance. A very sensitive balance was used, which, after having been counterbalanced for the weight of the tactual surface, tipped when the imbalance was as little as .1 g. The balance plate on which the tactual surfaces were laid was arranged so as to allow freedom of vertical movement, with any lateral movement excluded. Thus, the tactual surface could not be pushed aside laterally by the touching

finger. With the scale counterbalanced, the tactual surface gave way even to an extremely light contact, which was scarcely noticeable to the finger. The greater the counterweight placed on the free balance plate, the more forceful the touch had to be before the tactual surface withdrew downward. By varying the counterweight, any particular intensity of pressure for touching could be achieved. The subject was instructed to touch with one finger so as to maintain contact with the tactual surface as long as possible, and to avoid violent, jerky movements of touch.

I will briefly state the results of the experiment. The less pressure the subject exerts, the fewer materials that are recognized. Thus, decreasing the pressure has a reducing effect. At the lowest level of pressure, all materials were unrecognized, except for the fur, which was betrayed by the tickling effect of its hairs, and the glass and metal, where the material could be deduced from the sensation of smoothness; contact was too light to allow the heat flow from the touch organ to the tactual surface to contribute normally to recognition of the materials (Section 36). With the lightest contact, differences in hardness and softness could no longer be perceived, the spatial sense of the skin could no longer discern the coarse texture of the materials, and the vibrations necessary for the recognition of fine texture were lost as a result of the minimal friction.

Footnotes

1. I considered it permissible to include the latter times in the mean, because they, too, reflect how difficult the task was.
2. H. von Helmholtz, *Die Tatsachen in der Wahrnehmung* (Facts in perception). Berlin, 1879, p. 19.
3. As already mentioned previously, however, the reverse holds as well.

Section 30. Experiments with Amputees

I must insert here a brief summary of the results that I obtained in experiments with amputees on the recognition of materials (Katz, 1921, Section 17). Twelve of the materials listed above and an additional one (corrugated paper) were used. With his eyes closed and ears stopped, the amputee had to touch the surfaces with his stump, using whatever type of touching movement he chose; the tactual performance of the amputee's intact hand was used for comparison. The tactual surfaces were presented in a different order to the hand, which nevertheless had an advantage as a result of the single previous presentation of the tactual surfaces on the stump. The experiments involved 19 forearm amputees (F) and 16 upper arm amputees (U). Subjective tactual equivalences were numerous here, too, but since they revealed nothing new as compared with those previously presented, they need not be discussed further. Inappropriate judgments were accepted as correct in the case of tactual equivalents. The mean number of correct and incorrect cases are as follows for all amputees:

	Correct	*Incorrect*
F hand	9.7	3.3
F stump	6.7	6.3
U hand	9.6	3.4
U stump	6.2	6.8

The fact that more than half of all tactual surfaces touched by the stump were correctly discerned for F, and nearly half for U, indicates a considerable capability for the stump. The superiority of the hand over the stump in recognizing materials is not as great as one initially would have expected. If these tests show a considerable tactual capability of the stump, then it is understandable why experienced amputees speak of an impoverishment of the stump in sensation when an artificial limb is attached.

How does the amputee touch with the stump? He prefers to use that part of the skin surface that has a firmer character. Often it is the portion that is located directly above the bone of the stump. This portion provides better resonance for vibrations. In endeavoring to recognize the material, the amputee alternates between exerting pressure on the material (recognition of softness, deeper character) and moving the skin surface over the material (recognition of superficial character). The movements made are sometimes slow, sometimes fast; the latter seem particularly important for the formation and verification of hypotheses.

How are the experimental results to be explained? We seldom touch with the portions of the forearm and upper arm used by the amputees, so those must be regarded as positions for which normal persons have had little experience. This is not quite as true for the amputee, who endeavors to get as much benefit as possible from the stump in his everyday activities, and therefore in touching as well. Thus, a certain amount of practice contributed to the favorable results of our experiments. However, this involved more general or formal than specific practice. Everyday experience would not have brought the amputee into contact with most of the materials that were used. For this reason, I rather think that the statements made in Section 15 on the slight effect of specific experience on the tactual feats discussed there would hold here as well.

When an amputee is given the opportunity to practice touching the materials and to follow the touching with his eyes, he quickly learns to correctly designate tactual surfaces that he earlier had confused. In this way, then, linguistic designations are attached to small tactual differences that previously had not been noticed at all. There are individuals born armless who learn to use their feet with an astonishing degree of agility. The best known such case is that of the foot artist Untan [Katz, 1930]. It is not surprising, given what was said above about touching with it, that the large toe also seems especially disposed to improve with practice in the touching of the foot artist.

In order to get some experience in touching with the toe on new materials, I did the following experiments. I had an assistant present the tactual surfaces described in the preceding paragraphs in random order to the large toe of my left foot for touching. To my own amazement, the individual tactual surfaces were recognized in most cases by far. I do not know when I ever used my toe to touch glass, oilcloth, glazed paper, and still other materials, which now were readily recognized. Thus, the toe could hardly have had specific experience.[1] What we said above (Section 26, Subsection 2) about the recognition of papers by the toe has now been confirmed for a large number of very different materials. The theoretical interpretation of the results therefore will be very similar here as there.

In many cases, tactual recognition of objects (stereognosis) depends on, or at least is facilitated by, the recognition of the material. The material is so distinctive for many objects that we are spared having to make a more precise apprehension of the form of the object. Naturally, one then may ponder whether pathological astereognosis is due more to a failure in the recognition of the substance or of the form of the object. A separation of the two factors is possible through the use of intermediaries (gloves) that preclude, or at least considerably attenuate, the recognition of the material of the object, but not the recognition of its form. If an object has a very distinctive form, then it will be recognized even when the recognition of its material is precluded.[2]

Footnotes

1. Editor: Katz's suggestion that the foot is rarely used for texture judgments is disputable. M. A. Heller (personal communication, October 1988) said: "I think that we constantly make these texture judgments, but rarely verbalize them (unless we fall on the ice). While it is true that we don't normally use the foot in a laboratory, we constantly use the foot for texture judgments in everyday life. We instantly notice when we step on a slippery surface (. . . or we fall). When driving, I immediately know when the car hydroplanes (via the gas pedal). I doubt that walking could proceed normally if the foot (the glabrous skin) were subjected to anesthesia."

2. In these cases, as in those H. Hoffmann encountered, stereognosis can naturally be quite independent of the reduced pressure sensations. *Stereognostische Versuche* (Stereognostic experiments). Dissertation, Strassburg, 1883, p. 149.

Editor's Notes on Chapter II, Division II: Studies on Touch Transparency (Sections 31 and 32)

In the next chapter, Katz's focus shifts from the two-dimensional surface and its phenomenal properties, covered in Chapter I (Section 18 to 30), to the three-dimensional volume and its phenomenal properties. In Section 31, he returns to the phenomenon of touch-transparent film, which he covered in Section 8. He studies the discrimination of differences in the thicknesses between papers, which were felt between the thumb and index finger. There are problems with the study, however. Katz's inability to manipulate the thickness variable (he had to use commercially available papers) precluded obtaining precise measurements of the difference threshold. Also, the relative value of the difference threshold (Weber's fraction) did not remain as constant as Katz indicated it did. Notwithstanding these problems, Katz's phenomenological approach was quite successful in this case. He discovered that as thickness increased, the basis for its discrimination went through three stages. Only with the thickest sheets did subjects actually discriminate thickness on the basis of distance between the fingers. The thinnest papers were discriminated on the basis of the veiling effect of the paper placed between the fingers. Moderately thick papers were discriminated on the basis of their flexibility or bendability. He also found that movement of the thumb away from the body is more important than that towards the body (see Section 33 for a seemingly contrary finding).

In Section 32, Katz returns to the phenomenon of volume touch, which he covered in Section 7. He makes the point that volume touch vanishes, and the object underneath the material also is not experienced, if the intervening material is too hard. He also describes cases in which the blind chose to read braille through an intervening material, such as an apron (see also Section 24). This anticipates modern work on touching through cloths (see Editor's Introduction), but Lederman (personal communication, June 1988) said that the blind may choose to read through an intervening material not because it makes the characters clearer, but because it reduces the unpleasant sensations felt with continued touching.

Chapter II: Studies on Touch Transparency

Section 31. Sensitivity to Thickness

The feats of the touching hand reported on in this section somewhat surprised me at first. I learned that the hand discriminates differences in thickness of thin, flexible sheets with astonishing precision.[1] Touching was done by placing the thumb and index finger opposite each other and then inserting the paper to be tested between them. The apparatus we thus improvised corresponds to the calipers used by technicians to measure thickness. We have here a special case of the touch-transparent films described in Section 8, with the touch transparency occurring from both sides, rather than from one side, as in that case.

The first task that had to be done was a technical one. It consisted of assembling a large set of sheets that were as similar as possible in material and in surface structure, but as different as possible (and by approximately equal amounts) in thickness. I examined perhaps 200 papers in order to obtain a selection that met these requirements to some extent. The most commonly available calipers used for measuring papers and sheet metal are accurate to .01 mm, and somewhat reliable to .0025 mm. It soon became apparent that this precision was not sufficient for our purposes in devising a suitable set of stimuli. I therefore measured the thickness of the papers with a small device constructed by the Goettingen School for Precision Tool Making for their own use, mainly for measuring the thickness of screws. This small precision instrument allowed a direct reading of thickness to .001 mm. The device need not be described in detail, except to note that

145

the moveable part that was critical for the measurement always set down on the paper sheet to be measured precisely the same fraction of its own weight. Each paper was measured several times at different positions, but the deviation from one position to another usually amounted to only a few thousandths of a millimeter.

The thinnest paper used had a thickness of .027 mm,[2] followed next, in order, by .028, .031, .032, .033, .038, .043, and .048 mm; papers of up to about .048 mm are customarily called tissue paper. At .055 mm, the extremely thin writing papers began, followed by ten progressively heavier papers, up to the one at .093 mm, which resembles the paper commercially designated as "3A standard paper." Otherwise, it need only be mentioned that one paper in the set used was of the type of official postcard issued in 1919, with a thickness of .125 mm. The definitely cardboard-like papers began at .185 mm. The whole series of 30 papers extended from .027 to .302 mm. Since we could not fully control the gradation of the "thickness stimulus," but had to be content with the series of stimuli once it had been devised, experiments adhering to the strictest rules of psychophysical methodology were precluded. Future research, should the need for it be shown, may attempt to produce a more suitable set of stimulus materials and thereby determine more exact threshold values; we will content ourselves with less systematic experiments, which are yet informative in their own right.

We chose one paper as a "standard paper," and determined the nearest paper that seemed to the subject to be clearly thicker. The second paper had to be judged *thicker* three times in succession, without fail. As can be seen, we have placed strict requirements on the reliability of the discrimination. The discriminations reported here cannot be regarded as optimal, however, because a comparison paper closer to the standard paper than any in the present set might yet have received a "clearly thicker" rating. As a rule, we worked with three different comparison papers, each of which was presented three times in random order.

The subject was free to decide how to touch each paper and for how long; the freedom was not misused, because the touching time for a paper rarely exceeded 10 sec. The subject was supposed to act as if he or she had to test the thickness of the papers before buying them. In such a situation, one knows quite well how long it makes sense to continue touching. The particular touching movements of the subject cannot be described, but the attitude underlying them is completely clear, and anyone who should choose to repeat our experiments will obtain it.

A few more words on our method. One must distinguish between such observations as, for example, a just noticeable difference, whose magnitude is well specified by the strict rules of psychological methodology, and the testing procedure customarily used to determine a practically meaningful difference. For practical purposes, one will rarely have any use for a value determined in an absolutely flawless, scientific manner, such as, for example, a threshold value. Take the thickness threshold determined in that way: it would sometimes be recognizable, sometimes not. A thickness difference that has been found to be noticeable in practical matters has much more chance of being recognized as such time and again, given the same attitude.

With his eyes shut, the subject handles the papers passed to him one after the other by the experimenter. Some subjects use only their thumb and index finger, others their thumb, index, and middle fingers. The thumb opposes the other fingers. Either the thumb moves more, or the other fingers move more. The fingers never stay still. The thumb generally moves more rapidly and with heavier pressure away from than towards the wrist. Movement of the thumb away from the wrist is evidently more important in making a judgment. With repeated to and fro movement, the movement of the thumb towards the wrist only serves to make possible the reverse movement, which is the one actually intended. This is somewhat reminiscent of the process in sniffing, where testing of the air occurs at inspiration, but expiration is necessary in order to make inspiration possible. Just as one can be "all ears," so in this case one can be "all hands." One is completely tuned to the object, paying not the slightest heed to the touching movements, and therefore has not the slightest idea as to how many times the fingers have moved to and fro, which fingers participated and in what manner, and how long one has touched. The entire action up to the rendering of the judgment represents one unit, held together by the orientation of the task. Further analysis reveals three different periods: 1) a rather superficial contact with the paper, 2) the formation of the judgment, and 3) continued touching to check its correctness.

As observers in the longer series of tests, we used Mr. Dittmers and Miss Balcke, both philosophy students, and my wife; several other subjects served as well for occasional observations. The sensitivity to thickness changed quite considerably with the absolute thickness of the papers. With the thinnest papers from .027 to .048 mm, all subjects discriminated, virtually without exception, papers that differed by .02 mm, and even .01-mm differences were recognized by many with surprising reliability! With papers that are like writing paper, sensitivity is not as great, but still quite considerable. With cardboard-like paper of .125-mm thickness, for example, differences of about .04 mm can still be recognized. If one goes from this type of cardboard, which still bends to a certain extent under weak pressure, to cardboard and pasteboard that no longer bend even under heavy pressure, then the difference threshold jumps quite considerably. A 1-mm thick cardboard can only be discriminated from a second one that is approximately 1.3-mm thick. Although the relative difference thresholds vary little for sheets of different thickness, I hardly need say that the experimental results cannot be explained by Weber's Law, because completely different judgmental factors come into play when the absolute thickness changes.

With the thinnest papers, which snuggle completely against the touch organs when pressed from both sides, the degree of veiling through which the opposing fingers feel each other is critical for the judgment of thickness of individual papers and the judgment of thickness differences. We discussed above the tactual complex produced when two or more fingers of the same hand touch each other. When the paper membrane is inserted between the fingers, there is a fading of the tactual complex, which varies with the thickness of the paper. Actually, then, what is really compared is not paper thickness, but the degree of fading of the tactual complex produced by the different papers presented. We merely [operationally] define the judgment as one involving the thickness of paper. While the touching fingers can still feel each other's epidermal ridges through the thinnest papers, that is no longer so with the cardboard-like papers. Judgments in that case are not yet based directly on the feeling of distance between the touching fingertips, but on differences in the flexibility of the paper. Instead of a unitary tactual complex of fingers rubbing against themselves, we have two surface touches, which refer to the two sides of one object. At this second level, a completely different sensitivity to differences, one dependent on their absolute magnitude, emerges. A third level occurs when the thickness of the cardboard no longer allows any bending. Here, where the distance of the touching fingers from each other becomes really critical, the sensitivity to thickness differences, measured by their absolute values, decreases precipitously. From this point on, one could pass without further discontinuities, into testing the sensitivity for considerably larger thicknesses, which finally requires the use of two hands ("when the upper surface of a table is touched with a finger of one hand, and the lower surface with a finger of the other hand, then one is able to tell, with the eyes shut, just how thick the table is,"[3] Weber, op. cit., p. 89), but we would thereby enter an area we had not intended to investigate.

I also did some tests using sheets of gelatin, which is thoroughly unlike paper in its surface qualities as well as its elasticity. With a .055-mm thick gelatin, I found essentially the same sensitivity to thickness as with an equally heavy paper, despite the difference in material.

How do the eyes fare in comparison with the tactual sensitivity to thickness just investigated? There is an extraordinary superiority of the sense of touch for the thinnest papers. With the eye, for example, .027-mm and .048-mm thick paper, with sharply cut edges (great care must be taken in this respect), cannot be discriminated from each other. The thickness cannot be seen at all in such papers, but can be seen with cardboard at a thickness of about .2 mm.[4] Even if the viewing conditions are selected so as to be as favorable as possible, the edge of the thinnest papers yet remains visually extentless (Rubin); we see it merely as a contour that circumscribes the paper and demarcates it from the surroundings. On all papers whose edge appears extentless to the eye, the sense of touch far surpasses the eye in sensitivity to differences in thickness. We have again encountered a domain where the micro-level capability of the sense of touch overshadows that of the eye. In going from the thinnest papers to thicker ones, then we soon reach a point where even the eye can discriminate the breadths of the edges of the papers, but as long as the fingers are able to rely on the flexibility of the sheets, they yet surpass the eye in sensitivity. With heavier cardboard, where the judgment really must be based on how far apart the fingers are, then the eye surpasses the sense of touch.

Footnotes

1. On the great sensitivity of the blind to differences in thickness, see S. Heller, *Philosophischen Studien*, *11*, 1895, p. 421 f.
2. By comparison, the thickness of a thread from the cocoon of the silkworm is .005 to .01 mm. H. Henning noted in this regard: "I saw an East Asian weaving, which was about a cubic centimeter in size when folded together, but which produced a balloon 5 meters in diameter when blown up." Experimente an einem telekinetischen Medium (Experiments on a telekinetic medium), *Zeitschr. f. Psychol.*, *94*, 1924, p. 284. From this statement, one would calculate the thickness of the fabric to be approximately .000014 mm. Plainly, something like that is impossible.
3. In gynecological practice, this type of touching plays an important role, since both hands are used in examinations during pregnancy to obtain information about the position and characteristics of the fetus.
4. We will completely disregard knowledge of thickness coming from a different visual source, namely, transparency; this source, after all, is often misleading.

Section 32. Touch-Transparency of Space

On a visit to a home for the blind in Neukloster, a teacher there told me that blind girls who are supposed to recite memorized material try to make this task easier for themselves by placing the braille text under an apron and then reading it through the apron.[1] The practiced fingers can recognize the relatively pronounced elevations of the blind alphabet rather easily through the thin woven fabric. The blind pupil is hardly aware that a space is thereby touched through. As already mentioned above, the experience of the touch-transparent space of the body holds less interest for the palpating internist than the organs that he touches through this space. However, should he so choose, he can turn his attention to the space, contrary to his well-ingrained disposition. When experiments corresponding to the palpation procedure of the physician are conducted in the psychological laboratory, then the subject usually has two experiences, side by side, that of touch-transparent space and that of the touched object. What was already said above in Section 7 about the touch-transparency of space will be somewhat supplemented here with respect to the genesis of those impressions.

To provide the experience of volume touch, the material used (blankets, cotton, cellulose, cushions) must be sufficiently compressible. If it is too hard, or becomes so hard under pressure that the touch organ receives no increase in the pressure sensation with very heavy pressure on the object lying underneath, then neither this object nor the volume above it are experienced. A cushion filled with down works as a rule, but not one filled with horse hair. Touch transparency also vanishes when the touch organ is too soft to compress the material sufficiently. "Deep palpation is impeded when the touch organ is very much softer and more compressible than the tissue to be felt" (Goldscheider and Hoefer). Hausmann pointed out the surprising, but easily confirmed, fact that the tongue cannot be used to feel the radial pulse or to find the artery.[2] The tip of the tongue is simply too soft to compress the tissue lying above the artery. One can only agree with Goldscheider and Hoefer when they summarize their experiments on palpation as follows: "Only when the skin of the touching hand has been so compressed that its further compressibility is equivalent to that of the tissue layer to be felt, does continued penetration into depth against the resistance situated there evoke the requisite increase in the pressure sensation (op cit., p. 316)."

Only very small objects can be touched with a hand through a space. Unless expressly prohibited, a person will always touch with both hands; this makes the experience of the touch-transparent space clearer. It is obvious that visual elements come into play here. This is probably always the case even for palpation with two hands.

Footnotes

1. L. Cohn, who became blind at the age of 6, reports that he could read braille text well with cloth or suede gloves, and very well through kid leather gloves. Beitraege zur Blindenpsychologie (Psychology of the blind). *Zeitschr. f. angew. Psychol.*, Beiheft 16 (Supplement 16), 1917.
2. T. Hausmann, *Pfluegers Archiv*, *194*, 1922, p. 623.

Editor's Notes on Chapter III, Division II: Analysis of Touch Movements (Sections 33 to 35)

In the next chapter, Katz describes the ingenious methods he devised to graphically record the direction and extent of finger movements. In Section 33, a writing device attached by string to the touching finger was used to record that finger's movement on a kymograph (soot-covered moving drum). In Section 34, subjects touched briefly on a soot-covered paper with several fingers. In Section 35, Katz pressed his left index finger down on the soot-covered sheet with varying pressure, in order to show how the flattening of the fingertip increases with pressure.

In Section 33, Katz reports that lateral movement of the finger towards the body is more important than that away from the body. He does not explain this finding; it may reflect differences in how the finger digs into the tactual surfaces, or differences in central attention during these movements. In Section 31, Katz reported a contrary finding, which again he did not explain. When a paper is held between the thumb and the index finger in order to judge its thickness, movement of the thumb away from the body is more important than that towards the body.

In Section 34, Katz reports that a complete, continuous surface is felt even though the finger movements, as revealed by the soot tracks, cover only fragments of the surface. The effect corresponds to the filling in of the blind spot in vision, and it persists even when subjects plainly see the unstreaked portions left between the fingers. Katz refers to a central mechanism that reconstitutes the missing parts, but offers no further explanation of the effect.

Chapter III: Analysis of Touch Movements

Section 33. Graphically Recording
the Movement of a Single Finger

The following apparatus (see Figure 2) served to graphically record the movement a single finger made in touching a surface. A writing device was attached in front of the soot-covered drum of a kymograph. One lever bore a writing tip made of paper, and attached to the second, equally long lever was a long silk thread, which led to the finger whose movements were to be recorded. It was found that normal movements were almost purely lateral for the subjects and the particular experimental conditions used here. Therefore, the thread was fed over a roller located at finger level just at the side of the finger. If longitudinal movements had been more natural, we would have fastened the roller in that direction. This type of fastening and guiding of the thread was designed so that a finger movement produced an equally large movement on the lever. Since the recording arm was just as large as the lever arm to which the thread was fastened, movement of the finger was recorded directly in its natural scale on the drum. At a certain tension on the silk thread (i.e., at a certain motionless position of the touching finger), a weak rubber band fastened on the double lever gave the lever an exactly horizontal position. Any change in speed or direction of movement was revealed in the recorded curve.

After the thread was looped around the finger of the comfortably-seated subject, the hand, which had been resting on a cardboard, was so positioned by the experimenter that the recording lever was precisely horizontal. Papers from the basic experiment were presented for touching and qualita-

Figure 2.

tive judgment. The individual experiment was conducted as follows: the experimenter lifted the subject's finger a few millimeters off the pad, rapidly inserted the tactual surface, and then set the finger down vertically on the tactual surface. From previous experiments (Section 22), we know that this placement does not yet lead to recognition of the tactual surface. Immedi-

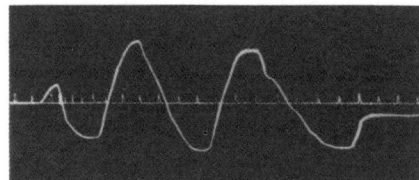
Figure 3. Subject Z.

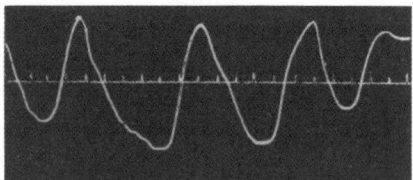
Figure 4. Subject L.

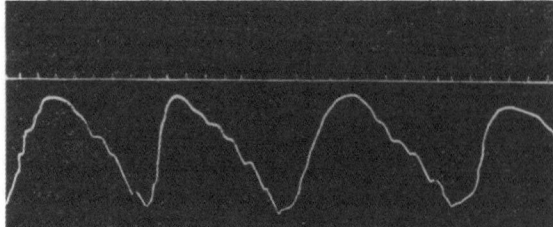
Figure 5. Subject B.

ately before placement of the finger, the experimenter signalled an assistant to start the kymograph. The subject then began freely touching until he thought he could give a correct qualitative judgment, whereupon the kymograph was stopped. Subjects were asked to avoid touching in too extensive a curve, and not to take too long to form a judgment; however, I do not believe that this distorted their behavior. In order to be able to compute the duration of the touching curves, the .2-sec cycles of a Jaquet chronometer were also recorded.

A few samples of the curves obtained are shown in Figures 3, 4, and 5. The first question that the curves ought to answer is the frequency of to and fro movements of the touching finger. Averaged across all tactual surfaces, this number (N) was 4 for Subjects Z and L, and 3.2 for Subject B (Buschmann). N varied considerably for the different tactual surfaces, ranging between 2 and 6 for Z, between 2.5 and 6 for L, and between 2 and 5 for B. Although the individual papers varied in ease of recognition across the three subjects, averaging the three values of N for each paper produced an order of means that seems to agree quite well with the difficulty in qualitatively judging the individual sheets. The more distinctive the tactual surface, the lower its N. The cloth paper (No. 14) indubitably made the most noticeable tactual impression, followed by that of the smoothest paper (No. 1).

Paper number	Mean N
14	2.2
1	2.8
2	3.0
4	3.5
10	3.5
11	3.5
9	3.8
13	3.8
7	4.2
5	4.3
3	4.5
8	4.5
6	4.7
12	5.7

The curves in Figures 3, 4, and 5 show that the touching finger never tarried, but moved to and fro all the time. For both L and B, touching curves always had a considerably steeper ascent than descent; that is, the movement was not carried out uniformly, but was more rapid toward than away from the body. Subject Z executed both movements about equally rapidly on the average. Are the movements in both directions equally important for the recognition process? When asked, all three observers reported, without the slightest deliberation, that movements toward the body were more important or, put more precisely, were completely decisive for the judgment. The tactual experience of the opposite movement was noticed very little, almost as though the movement away from the body was carried out only so that the movement toward it could occur.[1] The inward movement involved heavier rubbing on the tactual surface than did the outward movement. The mean speed (in cm/sec) of touching movement was calculated from the mean duration and extent of movement:

Direction	Mean speed (cm/sec)		
	Z	L	B
Inward	1.6	1.7	3.3
Outward	1.6	.9	1.6

We may assume that, based on instinct or past experience, the observers chose a speed that seemed optimal for recognition. For L and B, the movement toward the body is about twice as fast as that of the return movement, whereas for Z both were equally fast, but we will not pursue the individual differences any further. The speeds of movement that we calculated may appear slow, particularly in comparison with the findings reported above (Section 23), but it should be noted that the average speed was determined from the entire back and forth movement. If one takes the intermediate, steeply-rising portion of the curve, which is nearly linear, to indicate the *maximum* speed of the movement toward the body, then the speed is approximately 9 cm/sec for all three subjects. Only when one considers this—no longer inconsiderable— speed does the great difference between movement toward and away from the body emerge clearly. The observer seems to strive for this great speed in the intermediate portion of the approach movement. It is very probable that judgments are based primarily on tactual experiences obtained at the maximum speed.

Footnotes

1. On this point, compare what was said above (Section 31) on finger movements in touching for thickness. [There, movement of the thumb away from the body was more important than movement towards the body.]

Section 34. Movement of the Fingers in their Natural Configuration

The research described in the preceding section focused on the touching movements of the isolated, solitary finger. Clearly, the experimental conditions facing the subject there, above all the looping of a thread around the touching finger, could somewhat disrupt the naturalness of the movements, but being completely true to life in an experiment can almost never be combined with the methodological rigor of the laboratory. We strove to compensate as far as possible for the shortcomings of the preceding experiments by graphically recording the movement of the touching fingers in their natural configuration. The procedure was the simplest imaginable. We sooted glazed paper in the well-known way, and presented it to the subject for touching under conditions that will now be described. In touching, the fingers wiped away the soot, and the places that thus emerged as white revealed very clearly the path the touching fingers had taken.

Naturally, we could not allow the fingers to dance too long over the paper, because they would have wiped out their own traces while moving to and fro, ultimately leaving only a cleanly-wiped paper sheet for us. Accordingly, we had to limit ourselves to recording a characteristic phase of the movement by means of traces on soot. After preliminary tests, the following procedure proved expedient in this regard. The subject was told that it was up to him or her to rapidly form a judgment about the tactual surface presented. The experimenter grasped the subject's wrist, placed the fingers on the tactual surface, released the wrist, whereupon the touching movements began, and then lifted up the hand again. In order to acquaint subjects with this type of touching, as well as to learn whether tactual surfaces could be recognized with such a brief presentation, the experiment with sooted paper was preceded by a few tests using other tactual surfaces. It was found that subjects could obtain good information about the tactual surfaces with this procedure. Subjects had not the slightest inkling that a sooted sheet had been presented to them in the critical tests, as was shown by their astonishment about their blackened fingers when they opened their eyes. The judgment as smooth paper was always correct; the soot, which got onto the fingers when the touching began, did not noticeably impair the present tactual performance, but even if it had, it would not have mattered, because we were only concerned with analyzing the touching movement.

Four samples of the touching movement figures are shown, considerably reduced in scale, in Figures 6, 7, 8, and 9. The fingers are designated 1 to 5, from the thumb to the little finger. The figures reveal immediately that only Subject Z used all five fingers for touching, that Subjects L and philosophy student Krueger (Kg) both used three fingers, but not entirely the same ones, and that Subject K used four fingers. Each subject chose a different form of movement, K a hook-shaped movement, Z and L essentially a straight line, and Kg an elliptical movement. It may be briefly mentioned that Z chose the same direction of movement as in the experiments in Section 33 with the single finger, whereas L chose the direction perpendicular to that. However, we do not intend to resolve here such questions as whether an individual chooses the same form of movement time and again in touching, or whether there are typical forms of touching movement.

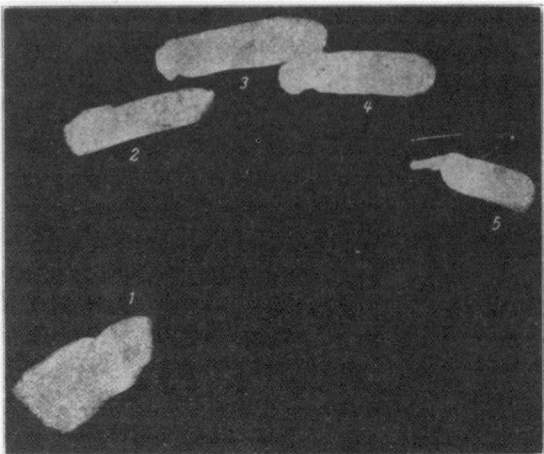

Figure 6 Subject Z.

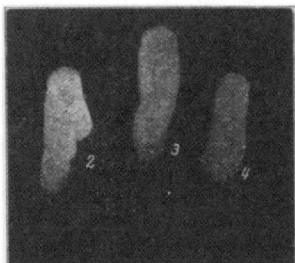

Figure 7. Subject L.

Of primary concern to me in this experiment was something entirely different. I call attention to the fact that all four subjects left completely unstreaked portions of the surface standing here and there between streaked portions. Thus, in normal touching, in order to perceive an unbroken sheet that is completely homogeneous from one touching finger to the next, the movement definitely need not occur in such a way that the sheet actually touched represents a self-contained whole, a continuum. I must again ask you to convince yourself of this fact by a tactual test of your own. One would be amazed if a visual check showed that the tactual surface consisted only of the fragments actually touched upon.[1] Obviously, therefore, a central mechanism reconstitutes and assimilates the tactual gaps to the fragments that were really touched.[2] Thus, we confirm here what was said above: the fingers work together as a team during touching and represent a unitary touch organ in consciousness. Even if one has already learned via a visual check that large tactual gaps are left in touching a surface with five fingers, the more critical attitude does not dispel the impression of a continuous,

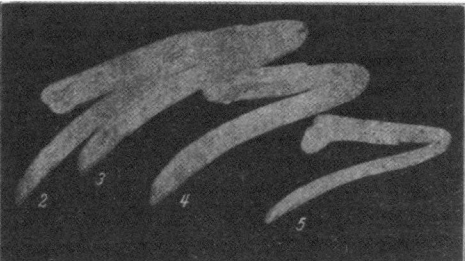

Figure 8. Subject K.

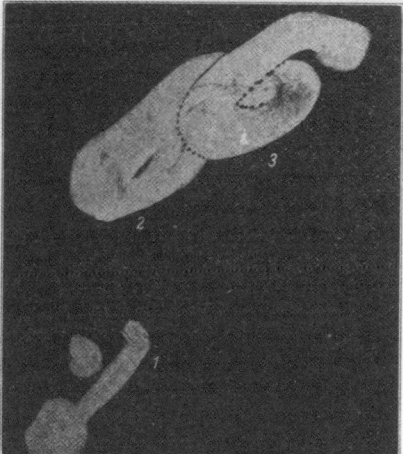

Figure 9. Subject Kg.

completely filled-in surface. I doubt that what I have just said pertains only to adult consciousness. Once the child uses the hand as a unitary grasping tool, it also becomes a unitary touching tool having the characteristic types of touching movements. Several facts, such as, e.g., the prevalence of the hand as a phantom limb in arm amputees (Katz, 1921), indicate that the hand as a whole is particularly well represented in the brain. Goldstein inferred from a pathological case he described, "that even the left hemisphere is involved in the touching of the left hand as well as the judgment of the movement sensations of the left hand."[3] Of the hand, it is again the fingertips that, according to Popper, have a particularly sensitive representation in the cortex.[4] In a case involving injury to the top of the skull, Popper found numbness at the fingertips on all contralateral fingers, particularly the four ulnar fingers, while the remaining sensitivity appeared to be essentially undisturbed.[5]

Footnotes

1. It hardly need be added that one does not doubt for a moment in touching with five fingers that one is dealing with a *single* object, in contrast to the two objects experienced with the Aristotelian touch illusion.
2. Steinberg says of the converging touch of the blind, "that in doing this, only the surface outline that the fingers glide along is directly apprehended; the surface itself is clearly restored in generative fashion during the movement" (op. cit., p. 143). Steinberg also reminds us here of the visual analogy involving the filling-in of the blind spot.
3. K. Goldstein, Ein Beitrag zur Lehre von der Bedeutung der Insel fuer die Sprache (A theoretical contribution on the significance of islands for language). *Archiv f. Psychiatrie, 55*, 1914.
4. E. Popper, Zur Organisation der sensiblen Rindenzentren (Organization of the cortical projection centers). *Zeitschr. f. d. ges. Neurol. u. Psychiatrie, 51*, 1919.
5. E. Popper, Beitrag zur kortikalen Lokalisation der Sensibilitaet (Contribution on cortical localization of sensitivity). *Neurol. Zentralbl., 37*, 1918.

Section 35. Deformation of the Fingertip in Touching

The extent of deformation of the fingertip depends upon the pressure exerted in a touching movement. When the pressure is strictly perpendicular to the tactual surface, a greater deformation means an increase in the area of the fingertip doing the touching. The soot method can provide certain information about the extent of this increase. A piece of sooted cardboard was placed on one balance plate of the scale used above (Section 29), and a variable counterweight on the other plate. My left index finger then exerts, in the most natural, unconstrained manner possible, just enough pressure on the sooted surface to balance out the counterweight on the other plate. I used 5, 50, 250, 1000, 2000, and 3000-g counterweights.

Figure 10 shows that the contact area of the fingertip increased 5 to 6-fold as pressure increased from 5 to 3000 g. The picture for 2000 g differs only slightly from that for 3000 g, because at such great pressure, the deformation of the fingertip has already about reached its outermost limit. As the pressure increases, the epidermal ridges undergo considerable flattening out, which can be seen clearly in the narrowing of the soot lines that remain standing. Finally, our photograph also reveals how the deformation of the fingertip shifts more and more from the distal to the proximal portions, as the pressure increases. Our illustration can only be properly understood if one conceives of it as engendered by the movement process at the touching finger. The *transformation with its individual phases*, which leads to the state of deformation depicted in the photograph, is that which characterizes the pressure experience.

QUANTITATIVE STUDIES OF THE TACTUAL PERFORMANCE 161

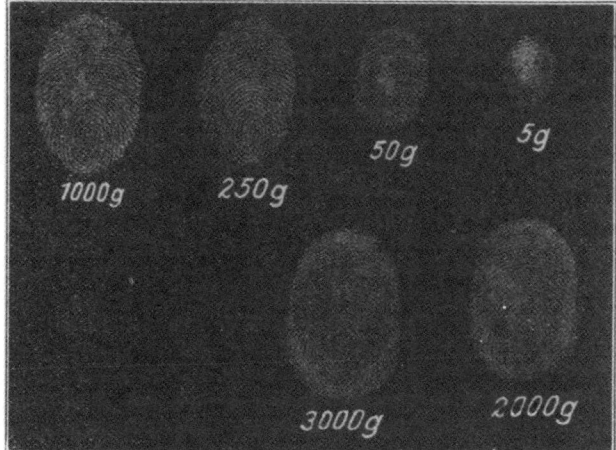

Figure 10.

Figure 11. 1-sheet   2-sheet   4-sheet

I have also taken a photograph showing the deformation of the fingertip from contact with elastic sheets (Figure 11). Two rods were fastened some distance apart on one plate of the scale, and very elastic, celluloid sheets were placed across them as a bridge. The subject had to exert just enough pressure on the center of the elastic bridge to compensate for the 200-g counterweight on the other plate. In order to vary the elasticity, 1, 2, or 4 celluloid sheets at a time were placed over each other. The pictures were produced this time by Subject K with printer's ink. The difference between the three photographs is not considerable, but the reduction in the deformation is clearly discernible in going from 1 to 4. What was said above about the interpretation of the pictures [Figure 10], holds here, too, where experiences of elasticity are involved to a greater extent; the tactual phenomenon can only be understood in terms of movement.

Editor's Notes on Chapter IV, Division II: The Role of the Temperature Sensation in Touch (Sections 36 to 38)

In the next chapter, Katz presents some of his most interesting ideas and experiments. This may well be the best chapter in the volume, yet the work has evoked few follow-up studies by other investigators. Katz ranked the temperature sense ahead of the pressure sense, but behind the vibration sense, which will be covered in Division III.

In Section 36, Katz shows that under normal conditions, thermal properties (thermal conductivity, thermal capacity, specific heat) produce good recognition of materials. (The specific heat of a body is the heat absorbed or given off per unit mass per unit change in temperature. The thermal capacity of a body is the heat absorbed or given off per unit change in temperature. Thus, thermal capacity equals mass times specific heat.) He then very cleverly demonstrates our dependence on thermal properties by using atypical conditions in some tests. Heating the materials just up to body temperature made them much less discriminable, for example, whereas heating the materials above the body temperature reversed their cold-warm ordering. In Section 37, he shows that presenting the materials in an unusual size, e.g., as tinfoil, also distorted the thermal gestalt. Since a material that is a good thermal insulator can be perceived behind a thin layer of a material that is a good thermal conductor (Section 37), the temperature sense is in part a remote sense, he notes. A buffer medium that transmits vibrations, but not heat, destroys the liveliness of the touch impression. The resulting stiff and dead feeling may explain why amputees often prefer to touch things directly with the bare stump rather than by means of an artificial device, Katz notes (Section 36). In the Addendum to Section 36, he points out that at various temperatures, perspiration wetness contributes to the discrimination of such materials as metal and glass (see also Katz, 1930).

In Section 38, Katz presents thermal analogs to the immersed, surface, and volume touches covered in Sections 6 and 7. These might be termed immersed warmth, surface warmth, and volume warmth, although Katz explicitly uses only the last term. Volume warmth occurs, he notes, when the head is turned in front of a hot oven, thus localizing the heat dome. The greater skill of blind persons at localizing objects through the temperature sense (e.g., they feel the approach of a warm cylinder at three times the distance that a sighted person does), suggested to Katz that such a heightened sensitivity very probably participates in the so-called sixth sense of the blind.

A temperature sensation generally favors the subjective pole, according to Katz, and warmth and cold may be experienced as pure states of one's body. The latter inclination may be overridden by movement of the body part, however. Also, the temperature impression normally is not alone, but is closely bound up with an impression of touch, especially on the fingers and inner hand and when the temperature is not too high or low. The tactile system dominates the thermal system when the two are closely bound together, and temperature significantly affects the perception of roughness (Green, Lederman, & Stevens, 1979). The bond between touch and temperature is weaker on parts of the body used less frequently for touching. As a result, the tendency for coldness to increase perceived heaviness is stronger on those parts of the body (Footnote 8, Section 38). The subjective pole emerges strongly when the temperature sensation is separated from the dominating embrace of the touch or tactile sensation.

The subjective pole also is favored if the fingers are warmed or cooled by being dipped in water rather than by touching a solid object. Thus, immersed warmth, like immersed touch, weakens the objective pole. Katz explains this weakening in an interesting way. In moving the immersed hand through a liquid or the air, one still perceives the coldness or warmness as residing in the object "out there" (objective pole), but it is perceived as a (temporary) state of the object, not as a (permanent) property. In the same way perhaps, the dark film of a shadow cast on a wall is not seen as being a part of, or properly representing, the true color of a wall.

Chapter IV:
The Role of the Temperature Sensation in Touch

Section 36. Thermal Qualities of Objects at Varied Temperatures

1. *Visual reflectance (albedo) and the thermal character of materials.* From my first observations on, it was clear to me that the temperature sensations accompanying and permeating tactual impressions influence considerably, if not decisively, the recognition of many materials by touch. Metals, like glass, betray themselves by their distinct coldness, whereas woollen fabrics are especially distinct due to their pleasant warmth. These observations were confirmed, in systematic experiments, by how reliably those materials having the most pronounced temperature impression, that is, metals and glass, were identified. The experiments, which will be presented here, show that the temperature impression contributes to tactual judgments not only in those extreme cases in which it is particularly noticeable, but for almost every material. A glimpse of a new research domain opens up here, but one that can only be developed as a closely-related appendage to our previous efforts.

Diffusely-reflecting surfaces absorb and reflect a constant proportion of incident light. The proportion reflected is known in physics as the albedo of a surface. No matter how markedly the light striking the object may fluctuate in intensity, this quotient remains unchanged. As a result, the black-white rank order of achromatic objects remains invariant with level of illumination. The mechanisms of the eye, both physiological and psychological, that determine the final effect of the light stimuli need not concern us here; they leave untouched the basic fact of the invariance. The visual invariance may be compared with a similar, even if not quite as extensive, thermal invariance. Just as the visual invariance depends on the albedo, the thermal character of a material depends firstly, on its thermal conductivity, and secondly, on its specific heat. The thermal quality is but little affected by the room temperature, so the rank order of temperature impressions is, within limits, independent of the prevailing room temperature. The invariance is not as complete as in the visual area; its limits are determined by the temperature of our body.

2. *Experiments at normal temperature.* Objects lying sufficiently long in a room having a certain temperature, say 18 deg C, will all assume precisely this temperature. Why, then, in spite of this fact, do such objects produce quite different thermal impressions when touched? Such differences would be inconceivable if we were to sense the temperature itself of those materials. When a finger touches a cooler object in the room, the temperature disparity begins to be removed; heat flows from the skin to the object contacted.[1] To pick out the extremes, the thermal conductivity of metal greatly exceeds that of wool; thus, the finger conducts more heat to metal than to wool. This difference in heat flow must provide the basis for the different judgments of objects contacted, based on their temperature impressions. Our problem, however, is not yet completely solved. The thermal stimulation of the skin is referred to the object touched, and not to the person's own body. In addition, it seems remarkable enough that metal feels positively cold and wool positively warm, even though heat flows from the body to the object in *both* cases and wool can only heat up, at most, to the temperature of the skin. Since the touching organ must always release heat when it exceeds the external temperature, it would be reasonable to expect that all substances touched would feel cool to some extent, at least when the contact was brief. These questions and others will be examined in the following experiment.

We selected ten of the materials used above (Section 29), namely those listed under 2, 5, 13, 15, 16, 20, 21, 28, 35, and 40, adding as the eleventh a lead plate, and presented the items either in pairs for comparison, or individually for absolute judgment of their thermal impression. Both the fingers and the palm of the hand could be used in touching. Judgments were more reliable when the tactual surface was larger, presumably for two reasons. First, the heat flows faster from the skin at the initial contact, when the temperature disparity is at its peak, and the more new portions of the skin that enter into the action, the more frequently does the strong initial flow of heat occur. Second, the larger area has an advantage because of the summation of the excitation from the individual sensory areas. This is analogous to the improved recognizability of a color as its area increases. The pause after each trial was sufficiently long so that the subject's hand again assumed its normal temperature. In the main experiments, the observers were D, K, L, and Miss Kagan, a philosophy student, and, during occasional observations, R, Mr. Fischel (F), a medical student, and Mr. Noldt (N), a philosophy student. The paired comparisons produced the following series, based on all judgments (n = 4) of our subjects, and ordered from colder to warmer:

1. aluminum sheet metal
2. lead plate
3. glass
4. oilcloth
5. cardboard
6. wood
7. linen
8. silk
9. terrycloth
10. velvet
11. cloth

The variability of the judgments was extraordinarily small. Of the 16 materials originally tested, five were excluded because the differences in thermal quality were too small to produce reliable choices between items. In these initial experiments, it was important to work with a series whose members could be readily discriminated thermally, and such a series was obtained.

Values of thermal conductivity and specific heat can only be provided for the technically-defined materials. There is no reason to suppose, however, that the thermal characteristics of the other materials, if known, would fit our statement above concerning the grounds for the invariance of the thermal rank-order of materials.[2] In any case, the basic validity of our thesis would not be undone by relatively small deviations, which must actually be expected as a result of the special circumstances under which the materials were presented. The smoother and flatter the surface, holding other properties of the object constant, the greater will be the heat flow from the touching finger. Thus, an object having higher heat conductivity, but a rough and bumpy surface, may feel warmer than an object having lower thermal conductivity and a smooth, plain surface. In addition, the fact that the woven materials were presented on a pad having a different thermal character than the materials themselves must cause some deviation from the otherwise expected results; in Section 37 this deviation itself will be discussed.

The effect of thermal conductivity may be isolated by holding approximately constant the specific heat. If one compares the three materials listed in the following table, which have approximately the same specific heat, then the results do correspond to the difference in thermal conductivity, with sulfur feeling warmer than glass, and glass warmer than aluminum.

Material	*Thermal conductivity*	*Specific heat*
Aluminum	.48	.22
Glass	.001	.19
Sulfur	.0006	.18

The first experiment took place in a room at approximately 18 deg C. When it was repeated at temperatures of -3 deg to +26 deg C, the same ordering of the series of materials was found. Thus, invariance of the thermal series held with respect to variation in the room temperature within our chosen range. The basic significance of the principle—which, for brevity, we will call the *Thermal Invariance Principle*—is not diminished by the fact that at the lower temperatures the thermal differences were somewhat less definite and the end members of the series seemed to have moved closer to each other, thereby reducing somewhat the reliability of the judgments. It was unavoidable that the touch organ, the hand, would cool off somewhat in the air during the experiments at lower temperatures, which took place outdoors in the winter, and that it would cool off markedly and rapidly when contacting the colder materials, and this cooling off probably hurt the reliability of the judgments.[3] In later experiments, it will be shown that very high temperatures also impair the reliability of the judgments.

Let us now turn to new experiments. In one experiment, the touching no longer involved the hand but the lips. The subject brought the tactual surfaces to the lips in the way that seemed most advantageous to him or her. The same ranking of the materials resulted for the lips as for the hand, at temperatures of -3 deg to +26 deg C. The Thermal Invariance Principle thus is also valid for the lips. The thermal differences were clearer at the lips, and the judgments were made more rapidly and confidently. This probably was because the much thinner epidermis of the lips facilitates the outflow of heat. A very hot object can be grasped briefly with one's fingers without becoming unbearably painful, and the blacksmith even grasps glowing coals briefly. One could not do anything like that with the lips. The greater temperature sensitivity of the lips must help considerably on materials whose recognition is based in part on the temperature impression.

A second experiment obtained absolute judgments of the temperature impression. The tactual surfaces were presented in random order. We limit ourselves to a detailed account of the judgments given by Subject K: 1) aluminum: extraordinarily cold, coldness increases with the duration of touching; 2) lead: not as cold as aluminum; 3) glass: very cold; 4) oilcloth: definitely cool; 5) cardboard: neither cold nor warm; 6) wood: rather cool; 7) linen: neither warm nor cold; 8) silk: decidedly cool; 9) terrycloth: neutral to cool; 10) velvet: rather warm, no longer warm when touched for a longer period; 11) cloth: absolutely warm, without changing essentially over a period of time.

A rank ordering based on these individual judgments would not quite match the one obtained in the experiments with paired comparisons. The deviations, to be sure, involve only a few intermediate members of the two series, rather than any outlying members, and are not surprising when one considers that even in judging the ten members of a black-white series, a rank ordering based on absolute judgments would probably not be the same as one based on paired comparisons. A higher performance is demanded by an absolute judgment of sensory impressions than by a comparison. When one considers the unusualness of the task, the performance of Subject K, like the similar performances of the other subjects, has to be called very considerable.

3. *Experiments in a heat box.* If the temperature impression of an object actually depends on the heat flow from the skin, then completely new sorts of phenomena ought to arise as soon as the temperature of the object exceeds that of the touch organ. In this case, the heat must flow from the object to

the organism, and the greater the thermal conductivity of the object, the more readily this should occur. The order of temperature impressions produced by our materials ought therefore to reverse, so that aluminum feels the warmest and cloth the coldest. To test this prediction experimentally, the following heat box proved simple and expedient. Into a dark barrel with the front wall removed were placed four large bricks that could be brought to a high temperature in an oven. The space above the bricks could easily be raised to temperatures of up to 50 deg C. The tactual materials were inserted into this space, and also touched there by the hand. To prevent rapid cooling, the dark barrel was encased with a thick insulating woollen material. The front opening through which the subject inserted his or her hand also was covered so that the heated air could not escape. The tactual surfaces did not come into direct contact with the hot bricks; they were presented for touching on a thick cardboard that was attached above the bricks and separated from them by an insulating layer of air. Thus, the tactual surfaces were heated only by the air in the upper part of the surrounding dark barrel. During a sufficiently long stay in the air space, the tactual surfaces assumed its temperature, which was confirmed with a thermometer.

The enclosed air was raised to a standard temperature of 44 deg C, or approximately 7 deg C above body temperature. The materials, except for No. 2, were presented for comparison by pairs. The following rank ordering, from coldest to warmest, was obtained for Subjects K, F, N, and R:

1. cloth.   2. velvet.   3. wood.   4. oilcloth.   5. glass.   6. aluminum.
     (terrycloth) (cardboard)
     (silk)   (linen)

If we consider first only the six tactual surfaces not presented in parenthesis, then a comparison with the rank order presented above (Subsection 2) shows that the sequence has precisely reversed itself. Thus, we can say that at a temperature *above* that of the body, the Thermal Invariance Principle holds for a *reversal* of the usual thermal series. Thus, we confirm the supposition derived from our theory that temperatures which exceed that of the body must make the materials having a higher thermal conductivity appear warmer than those having a lower thermal conductivity. In no case was the supposition contradicted by a direct inversion of two materials. This series is less differentiated; it has become shorter; velvet, terrycloth, and silk, and, likewise, wood, cardboard, and linen can no longer be discriminated from each other. At this high temperature, all tactual surfaces feel uncomfortably hot, so that one does not like to let the fingers rest on them long enough to arrive at a confident judgment. This is not to say, however, that this alone explains the decrease in sensitivity to differences. The higher temperature not only eliminated many differences completely, but also lessened all differences of the series from 1 to 6; the end members, cloth and aluminum, stood much closer together than at lower temperatures. The shortening of the thermal series at low and very high temperatures reminds one of the shortening of the black-white series under very high and very low illumination. In both sensory systems, optimum performance occurs in an intermediate region, owing to the better adjustment of the sensory organs in this region.

I conducted a few experiments in the heat box at a temperature approximately equal to that of the touching hand. I could hardly discriminate metal thermally from cloth, plainly because the individual materials no longer produced different temperature impressions in the absence of a temperature gradient. Given the cessation of heat flow to the skin, it is very striking that the substances nonetheless all felt very hot to burning; we will return again to similar phenomena below (Section 37).

What happens to the absolute recognition of individual materials at high temperatures? It is diminished in all cases, but to very different degrees. Recognition suffers the most for materials that are normally quite cool, such as metal and glass. These two materials can no longer be recognized by their surface properties alone, and the same is true for oilcloth. All other materials in our series, those producing neutral or warm impressions, are indeed still judged correctly, though with considerable difficulty. Note well that we refer here to the absolute recognition of materials, not to the mutual discriminability of their surface structures; the latter is scarcely affected by changes in temperature. Thus, even metal, glass, and oilcloth are still discriminated from each other and from the other substances, only they are no longer recognized according to their material. What are the implications of these findings for normal temperatures? If we assume external temperatures below that of the body—which is definitely the prevailing state, after all—then the role of the thermal component in the tactual recognition of materials may be characterized as follows: 1) it is indispensable for those substances that, like metal, glass, and oilcloth, make a definitely cool impression; absolute recognition of such substances depends on their thermal properties as well as their surface properties; 2) it is not indispensable for, but merely facilitates the recognition of those materials producing a thermally-neutral or warm impression.

Other methods than those mentioned above have been used to study the significance of the thermal component for the recognition of materials. The following, for example, is very simple and makes use of the fact that the latency of sensory response is longer with the stimulation of warm and cold points than that of pressure points. We touch the material in our series so briefly that no heat transfer characteristic of the material can occur.[4] Then, all members acquire a thermally-neutral impression, or, more properly stated, an indefinite impression. What implications follow for the recognition of the materials? Recognition of the cool substances suffers much more than that of the warm materials. Metal and glass are not recognized, whereas wool is. The new method therefore produces the same results as were obtained by raising the temperature.

Likewise, interposing a good insulating material, e.g., a woollen glove, can almost completely eliminate the different thermal effects of the substances. If the recognition of most substances is then absent, this is, to be sure, due primarily not to the elimination of the thermal qualities, but to the change in the impression of the surface structure. When heat flow is prevented by a thermal insulator, the tactual impression suffers in what I should like to call its liveliness; it assumes a rigid, dead aspect. This becomes particularly clear, understandably, in touching a living being. I believe that amputees frequently prefer to touch with the bare stump because of the greater thermal liveliness of the impression, for which even the best prosthetic device, with its thermal insulation, can offer no substitute.

Why do definitely cool materials depend so much more heavily upon the temperature factor for their recognition than do thermally-neutral or warm materials? The answer, I believe, is that the impression of coldness, which is conveyed by very few materials, is much more characteristic of those materials than is the case with the impression of warmth, which most materials produce. Thus, the memory for cold materials is less dependent upon the nonthermal characteristics that can lead to recognition, i.e., those of the surface structure. The nonthermal characteristics are relied upon only to the extent that it is necessary to differentiate among the cold materials, which, along with glass and metal, include porcelain, brick, earthenware, marble, and plaster. If one of these cold materials is presented, then the memory for its surface structure is usually sufficient for its discrimination. If the thermal factor should be lacking in the overall impression, or if it should be reversed relative to the usual situation, then the basis for the discrimination would be removed. The situation is different for thermally-neutral or warm substances. They are not as impressive from a thermal viewpoint, yet they constitute the great majority of the materials we come into daily contact with. Therefore, we must pay more attention to the distinctive features of the surface structure and impress these more markedly into our memory, if we are to recognize the materials again later. Eliminating or markedly changing the thermal component does not prove fatal, since recognition can still be achieved, based on the impressed surface structures.

4. *Generic and distinctive thermal gestalten; preciseness* (Praegnanz) *of the temperature gestalt.* Every thermal impression that affects the character of a material has its own distinctive temporal gestalt. This has nothing to do with the slow arousal of the temperature sense, which has already been discussed, but, rather, it means that the temperature sensation that occurs

during touching lasts for a relatively long period (usually several seconds) and forms a unified whole.[5] The latter varies with the material and thus may characterize it thermally. We will speak here of a *temperature gestalt*. Every material has its own *distinctive* temperature gestalt, but there are groups of substances that are similar thermally, and whose temperature gestalten are similar to each other. Such is the case, for example, with the common metals that occur as objects in our environment, and with the common types of wood, fabric, and paper. Thus, there can be typical or *generic* temperature gestalten for metals, woods, fabrics, papers, etc. The temperature gestalten, having been impressed on the memory, aid in the recognition of materials.[6] More frequently, because it is easier, a generic rather than a distinctive temperature gestalt is recognized; thus, for example, one can more readily say, based on a thermal impression, that something is wood, rather than that it is oak or pine.

I carried out some experiments with a series of metals (copper, aluminum, brass, zinc, iron, tin, and lead) that can be discriminated from each other based on thermal conductivity and specific heat. All seven materials are recognized as metals on the basis of their generic temperature gestalten. As regards their distinctive thermal impressions, only a few can be reliably discriminated from each other, such as, for example, lead from copper; copper has approximately eleven times the thermal conductivity, and three times the specific heat, of lead. My experiments convinced me that one could, with practice, learn to discriminate almost all metals in our series from each other based on their distinctive temperature gestalten.

My own observations, as well as those of other subjects, indicate that in trying to recognize thermal properties by touch, one endeavors to educe the generic, and, if possible, the distinctive temperature gestalten of the materials as well, to the highest preciseness (*Praegnanz*) possible.[7] One first attempts to orient oneself by determining whether the material is warm or cold. If warm, one tends to execute touching movements within a small region; if cold, one makes more ample movements. Why? By touching in a narrow region, one heats up the substance, the heat flow is blocked by the poor thermal conductivity, and the impression that one is dealing with a warm object becomes even more definite. On the other hand, by constantly touching new points on a good thermal conductor, one accelerates the already excellent heat flow and thereby increases the impression that one is dealing with a cold object.[8] The particular type of touching therefore serves in both cases to bring out temperature gestalten of greater preciseness (*Praegnanz*). In the following paragraphs, we will discuss what else temperature gestalten can do.

Footnotes

1. "The temporal course of the temperature change at the skin surface, holding constant the contact area and the material, can be represented as a function of the amount of heat that must be transferred until the temperatures are equated." M. von Frey, in L. Tigerstedt, *Handbuch der physiologischen Methodik* (Handbook of physiological methodology), III, 1, p. 4. Leipzig, 1914.
2. Landolt-Boernstein, *Physikalisch-chemische Tabellen* (Physical-chemical tables), 3d ed., Berlin, 1905, and "Huette," 23d ed., 1920, give the following figures:

Material	Thermal conductivity	Specific heat
1. Aluminum sheet metal	.48	.22
2. Lead plate	.081	.031
3. Glass	.001	.19
4. Oilcloth	-	-
5. Cardboard	.00045	-
6. Wood	.00030	.65
7. Linen	-	-
8. Silk	.00006	-
9. Terrycloth	-	-
10. Velvet	-	-
11. Cloth	.000057	-

    These figures are scaled with reference to the thermal conductivity of silver = 1, and to the specific heat of water = 1. The value reported for wood gives the thermal conductivity when the direction of the cut is along the grain.
3. I cannot say whether the cooling off changed the thermal conductivity of the epidermis, which is well known to be very poor. The epidermis is bloodless, according to Spalteholz, whereas the uppermost layer of the corium (papillary layer) has a very good blood supply. *Arch. f. Anat. u. Physiol.*, 1893.
4. It is well known that a metal which is so hot that it would burn the skin with prolonged contact, produces no temperature sensation with a very brief contact.
5. A. Voigt also pointed out that in the case of the temperature sensation, the object of judgment is a process, not an immutable state. This author also provides a detailed theoretical account of the physical basis of heat transfer or equalization. *Zeitschr. f. Psychol.*, 56, 1910, p. 346.
6. Therefore, the absolute temperature memory is called into play. The capability of this memory must be considerable, according to our statements; instructive would be a comparison of it with the absolute color memory, which, according to the experiments by Lotte von Kries and Elisabeth von Schottelius, is also very good. *Zeitschr. f. Sinnesphysiol.*, 42, 1908.
7. The connection of the materials with their distinctive and generic temperature gestalten is purely empirical in nature. Bringing out the preciseness (*Praegnanz*) of a temperature gestalt means creating favorable psychophysiological conditions for its recognition. In Gestalt theory, the concept of "the preciseness (*Praegnanz*) of the gestalt" has a different meaning.

8. Even in slow cases, judgments of thermal quality do not exceed approximately ten seconds. Thus, there is no possibility that the temperature sensation will disappear completely as a result of a very prolonged contact; no complete thermal adaptation occurs. Holm found that with prolonged stimulation with the constant-temperature Thunberg temperator, the temperature sensation (on the abdominal skin) disappeared at 10, 15, 20, and 25 deg C, only after 165, 112, 72, and 47 seconds, respectively. *Skand. Arch. f. Physiol.,* 14, 1903. Thunberg's monograph on the skin sense is very instructive concerning thermal adaptation phenomena.

Addendum. Perspiration and Touching

The experiments at high temperatures can also help reveal what influence perspiration has on the tactual performance. A hand at normal temperature always shows slight traces of perspiration wetness. The amount of perspiration at the fingers can be determined subjectively somewhat reliably by touching the fingers together. Normal perspiration wetness can influence the tactual impression most markedly in the case of smooth, nonabsorbent surfaces that have good thermal conductivity, such as, e.g., metal and glass. The perspiration is deposited on surfaces of this sort and then acts as a slightly sticky coating. Everyone knows the halting, bumpy movement that occurs on glass and metal when one's fingers are perspiring profusely and that also tends to occur when the sweat glands are least active. In experiments at high temperatures in the heat box, the perspiration wetness on the tactual surfaces evaporates so rapidly that it can have no effect at all (assuming, of course, that the hand is exposed to the high temperature only briefly, and does not start to perspire profusely). The effect of evaporation in the heat box becomes most noticeable in the case of glass, aluminum, and oilcloth surfaces. This experiment thus clearly reveals the way in which perspiration normally affects the characteristic impressions of those materials. I rather think that the elimination of the influence of perspiration wetness on the tactual impression, along with the changed thermal impression, helps to make those materials unrecognizable at high temperatures. Most woven fabrics draw little perspiration unto themselves, because they are poor conductors of heat, and they are able to absorb the moisture they do draw, owing to their physical structure. Perspiration wetness has much less to do with determining the characteristics of such surfaces. That perspiration plays a role even for these surfaces can be seen from the way that the tactual impression changes when the touching fingers are supplied with a trace of drying powder. The surfaces then appear smoother due to the reduction in sliding friction. Other intermediate agents, such as collodion and adhesive tape, also prevent the perspiration wetness from having an effect during touching. Perspiration wetness tightens the bond between the touch organ and the tactual surface; heavy perspiration can increase it to the point of adhesive friction, which helps in handling heavy tools. In the case of calloused hands, where perspiration cannot develop its full effect, the worker handling a spade or axe assists it in a well-known but not very appetizing way.

In examining the functional equipment of the organism, it may be significant that the hand is probably the most heavily perspiring part of the body that is not covered with hair. For now, we only wish to emphasize it as one *fact* that perspiration on the hand is advantageous at least for handling many tools. Greater concern has recently been shown for the nervous mechanism of perspiration, which has considerable biological significance.[1] I have begun experiments to determine whether perspiration depends directly or indirectly upon the will.[2] The research also promises to throw new light on the more general significance of perspiration for touching. With a strong magnifying glass, one can follow quite well the emergence of beads of perspiration on the ridges of the papilla.

Footnotes

1. On this, see, e.g., L. Loehner, *Pfluegers Archiv, 122*, 1908.
2. Billigheimer (*Muench. med. Wochenschr., 68,* 1921) hypothesizes that scattered cerebral perspiration centers are connected to the individual cortical motor areas. The isolated increase in perspiration on the parts of the hand used for touching also points to such a system.

Section 37. Temperature Gestalt and the Thermal Capacity of Objects

The thermal capacity of an object depends upon the specific heat of its material as well as its mass.[1] We will conduct a few more experiments on the effect of the heat capacity of an object on its thermal impression. I attach a single sheet and an eight-sheet thick pile of thin tin foil next to each other on a thick cardboard pad. If I touch both surfaces by moving my fingers around, so that they do not tarry at any one position, then the surfaces appear thermally equivalent, that is, both feel cool. On the other hand, if I let my fingers remain longer in one area, then the eight-sheet stack provides a definitely cooler impression. The explanation is simple. The conductivity and specific heat are equal on both piles, but the thermal capacity of the eight sheets is greater, producing a heavier heat flow from the hand, and thus the impression of more intense coolness. The single sheet of tin foil is quickly heated by the fingers, allowing the build-up of heat in the more poorly conducting cardboard underneath to be clearly discernible in the temperature gestalt. The temperature gestalt takes an unexpected turn in the single-sheet case; the observer must reckon with a different sequel after the initial impression. Metal objects we encounter in daily life usually do not take the form of such a thin sheet as tin foil; they are less extended and more compact, and as a result can accept greater quantities of heat. Despite tremendous variation, owing both to choice and chance, in the distribution of various materials, a certain regularity nevertheless prevails with respect to the size of objects in our environment. Wood, as a rule, does not occur with the thickness of parchment paper. Wool, cotton, linen, silk, and all woven fabrics in general occur in the thicknesses peculiar to them. For other materials, such as paper, leather, glass, porcelain, and plaster, the thicknesses vary more widely, but not entirely irregularly. As a result, our practical experience concerning the thermal nature of the substances is not untouched by the prevailing size of their occurrence. Thus, the typical temperature gestalt of a substance also depends upon its thermal capacity. If a material takes a form that deviates *very markedly* from its usual one, then the temperature gestalt begins in its typical fashion during touching, but then pursues an aberrant course. When touching the single sheet of tin foil, for instance, the temperature gestalt begins as that typical for metal, but then bends off course as a result of the unexpectedly low thermal capacity.

From the experiment with tin foil, one can easily progress to other experiments that are equally convincing. Merely combine two materials that differ as much as possible in thermal conductivity in such a way that a very thin layer of one lies over a thicker layer of the other. After some practice, one can even succeed in describing the thermal properties of the hidden material based solely on the overall thermal impression. The temperature sense thereby assumes in this case the character of a remote sense.[2]

At this point, we return again briefly to the seemingly paradoxical fact that poor conductors of heat, such as wool, feel positively warm when touched by much warmer skin. Precise observation reveals that even with wool, the first temperature impression is actually that of coolness. Only after the material has been touched for some time, even if only for a few seconds, does it become positively warm. Hering has already made very valuable observations on this (*Sitz.-Ber. d. Wiener Akad. d. Wiss.*, *3*, Section 75, 1877), which Thunberg (op. cit., p. 684 f.) described as follows: "If we place a hand on a poor thermal conductor (e.g., oilcloth) at room temperature, then this feels cool initially, but the coolness soon disappears, giving way to a definite sensation of warmth, which to a certain extent increases and persists. In an analogous manner, we receive sensations of warmth when we put on gloves or clothes. The sensation of warmth that arises in this way is based on the higher surface temperature produced on the poor thermal conductor by the heat exchange."

In careful, informative studies on temperature sensations, Ebbecke extended Weber's and Hering's theory with his view "that the temperature sensation arises due to a difference between the temperature of the environment immediately adjacent to the touch organ and that normally maintained by the blood stream." However, when he infers from this "how subjective our temperature sensation is, and how the temperature sensation serves first to safeguard the individual, and only indirectly to provide knowledge of the physical processes,"[3] then I should like to point out, in contradiction to this, the statements in this and the preceding section that indicate the very considerable role the temperature sense has in the recognition of the physical properties of our environment. The temperature sense definitely occupies an intermediate position between the interoceptive and the exteroceptive senses, according to Sherrington's definition.[4] Its exteroceptive function has been underestimated. Our observations should lead to a fairer assessment of it.

Footnotes

1. "Each object that touches the skin and whose temperature differs from that of the outer layer of skin, thereby fulfilling the general condition for thermal stimulation, must also have at its temperature a certain minimum thermal capacity in order to produce any temperature sensation at all." Thunberg, op. cit., p. 684. Using small pieces of fine copper wire, L. F. Barker found that the smallest heat loss required for exciting the cold points was 2.4 microcalories. *Deutsche Zeitschr. f. Nervenheilk.*, 8.
2. An archaeologist once told me that by touching one can determine the points at which marble statues are covered with plaster. Presumably, temperature gestalten lead to the correct track in this case as well.
3. *Pfluegers Archiv*, *169*, 1917, p. 462 and p. 436.
4. This is also expressed in our language: "I am or I feel cold" or "I am freezing," and "it is cold" or "it is freezing." In the case of colors and sounds, our language provides no corresponding expression for what has been sensed. [Editor: Thus, according to Katz, one can say "it is red," but not "I am or I feel red." However, the latter expression can certainly be used metaphorically, as in "I feel blue." See also Footnote 9, p.43.]

Section 38. Fashioning of Temperature Impressions

1. *Localization and fashioning of pure temperature impressions.* Pressure points are so intermingled with hot and cold points on the skin that one must always reckon on simultaneous arousal of tactual and thermal points during surface stimulation. Indeed, as the statements made above make clear, all impressions received from materials have a special thermal coloration. The

tactual impression is impregnated and saturated with warmth or coldness. The obvious question then is whether the thermal quality also participates in the actual fashioning of the tactual impression. This question will be tested below.

In any case, there are pure temperature sensations that possess no trace of a form, and cannot be localized either. If one lies absolutely still in bed with very cold feet, then a painfully intense impression of coldness can exist which has no specifiable form and cannot be localized at some place on the body. Rather, this impression has a somewhat free-floating quality.[1] In addition, the impression of coldness is not projected onto something external; it has a purely subjective character, apparently lacking in the objective pole. The slightest movement of the feet produces localization; then, one merely has "cold feet." However, the sensation of coldness does not assume a fixed *form* when this occurs.

One can stimulate warm and cold points without any arousal of pressure points: the warm points with heat rays, and the cold points with falling drops of ether, which provide more point-like stimulation.[2] The localization of purely cold stimuli is actually possible, but is not quite as accurate as that of pressure stimuli, as indicated by several experiments.[3] This finding renders improbable any important influence of the cold sense on the localization of pressure sensations. Even more important is the observation made during these experiments that the pure impression of coldness produced by the drops of ether is referred to an external stimulator; it is already bipolar. With this, we have reached the first stage in the fashioning of a pure temperature impression.

If the face approaches a hot stove so that it is struck by the heat rays, then one has the impression of "directed" heat during the movement. The heat is referred to a remote source. If one moves away from the stove, then the impression of directed heat is not as strong. If the head is turned back and forth while maintaining a constant distance from the heat source, then its localization becomes more definite. However, the heat itself also develops a certain gestalt. It has a certain voluminousness, like a heat dome lying in front of one's face. Because the temperature gestalt is related to the volume touch described above, one could speak of a volume warmth.[4] This phenomenon is produced only by movement. The regular waxing and waning in the intensity of the warmth sensation on various portions of the face provides the impetus for giving form to the heat rays. I doubt that the slight head movements necessary in this case produce sufficient changes in air pressure to arouse the pressure sense. An immersed *coldness* sensation can be produced by moving one's head very close to an ice block, but it will not become as definite as the warmth sensation. We will learn more later about the role of visual qualities in the fashioning of these phenomena. Based on what has been said above about the fashioning of the pure warmth and coldness impressions, their participation in the tactual modes of appearance could indeed possibly occur only for volume touch. No such participation likely occurs, however, because the warmth and coldness completely lack the dynamic (resistant) quality that characterizes volume touch.

The localizing ability of the temperature sense, considerably heightened by practice, very probably plays some part in the so-called "sixth sense" of the blind. How enormously the sensitivity is increased in the blind is revealed by the fact that, as shown by Krogius' study,[5] they can discriminate whether the black or the white side of a heat cylinder filled with water at 42 deg C is turned toward them. Blind subjects detected the approach of the cylinder at three times the distance of sighted subjects.

2. *Fashioning of temperature sensations through pressure sensations.* Having shown above that the pure temperature sensation can do little on its own to achieve its own form, one is forced to assume that the thermal impressions that occur in connection with tactual impressions depend upon the tactual forms for their fashioning. If the point of the temperature probe touches us, then the form of the warmth or coldness impression is contingent upon the form of the point and the localization upon the preciseness of the localization of the point. The tactual impression takes the lead; it provides the framework onto which the temperature sensation is fitted. This holds true in the same way for both macromorphic and micromorphic tactual gestalten; down to the finest grain, the tactual impression is thermally impregnated.[6] As a result of this permeation, the objective pole of the tactual aspect strengthens the objective pole of the thermal aspect; the object that touches us acquires the properties of warmth and coldness, just as it possesses the properties of roughness and smoothness.

The connection between the thermal aspect and the tactual impression, which presumably develops through experience, depends on certain objective and subjective conditions.[7] Nowhere is the bond as strong as at the fingers and palm. The general rule seems to be that the bond loosens to the extent that the part of the body has been used less frequently for touching.[8] If a definitely warm or cold object touches us on the chest or the thigh, for example, then the two-tone combination of contact and temperature impressions is definitely not as indissoluble as on the fingers. As to the objective conditions for loosening the bond, only the transition to extremely high or extremely low temperatures should be mentioned. The subjective pole of the thermal impression emerges more strongly with its release from the tactual embrace under these circumstances.

When the fingers are dipped into a liquid, the temperature impression is referred to the liquid touched, but not quite in the same way as with solid objects. The thermal impression of water is interpreted more as a state or condition and less as a property or attribute, as with solid objects. Therefore, I will not speak of a distinctive temperature gestalt for water or other liquids.

Even less so than with liquids, where some may detect a slight step toward temperature gestalten, is there a tendency to interpret temperature impressions from the atmosphere as properties.[9] We usually refer to the temperature of the air, even if we are unadapted to it and thus sense it in a lively way, by saying: "*it* is warm," "*it* is cold."[10] However, if the air is moving, it takes on the form of an immersed touch, and we then perhaps speak of cold or warm *air* as well.

Footnotes

1. Ebbecke also finds that the cold sensations produced by immersing an extremity in cold water for an extended duration are "detached from localized individual sensations." *Pfluegers Archiv, 169,* 1917.
2. According to the procedure described by F. Kiesow, *Arch. f. d. ges. Psychol.,* 22, 1911. Kiesow stated that he obtained pressure sensations as well with this procedure, but von Frey spoke only of cold impulses (*Zeitschr. f. Biologie, 66,* p. 416). Fick, cited by Kiesow, said one cannot recognize whether circumscribed heat rays on the back produce a temperature or a pressure sensation. I believe that the simultaneous arousal of pressure points by drops of ether can be ruled out, because the sensation of coldness first occurs, as a result of evaporation, 1 to 2 seconds after the drop falls.
3. On this, see also Ebbecke, *Pfluegers Archiv, 169,* 1917, p. 453, and E. Gellhorn, Untersuchungen zur Physiologie der raeumlichen Tastempfindung... (Studies on the physiology of the spatial tactual sensation...), *Pfluegers Archiv, 189,* 1921. Experiments done by E. von Skramlik for J. von Kries showed "that an independent localization of the temperature sensations occurs, which frequently is no less accurate than that of the tactual sensation, but frequently is afflicted with singular errors." J. von Kries, *Allgemeine Sinnesphysiologie* (General sensory physiology), Leipzig, 1923, p. 17. As to the *two-point thresholds* of the temperature sense, von Skramlik himself stated that they "are of the same order of magnitude as in the case of touching, indeed at points they are even considerably lower." *Zeitschr. f. Sinnesphysiol., 56,* 1925, p. 134. M. Ponzo found that in general cold stimuli are localized better than, and warm stimuli equally as well as, tactual stimuli. *Arch. ital. de biol., 60,* 1913.
4. Based on these observations, I cannot entirely agree with H. Hoffmann (op. cit., p. 70), when he disputes the involvement of the temperature sense in the comprehension of spatial relationships.
5. A. Krogius, Zur Frage vom sechsten Sinn der Blinden (Question of the sixth sense of the blind), *Zeitschr. f. exp. Paedagog., 5,* 1907.
6. In the vicinity of an intense coldness (warmth) sensation, a thermally-neutral or weakly colored touch is sensed as coldness (warmth), according to Goldscheider. *Pfluegers Archiv, 165,* 1916, p. 21 f.
7. The necessary conditions for the integration of punctiform warmth and pressure have been studied by R. S. Malmud. *American Journal of Psychology, 32,* 1921, pp. 571-574.

8. In his research on Weber's illusion, M. von Frey found that the weight of cold objects was consistently and greatly overestimated on the forehead, but not in the palm of the hand. Von Frey hypothesized that the greater insistence of a cold weight as cold makes it appear heavier. This conception fits in well with the view presented here that the connection between coldness and heaviness is looser on the forehead than on the hand. As a result, the coldness impression can more readily influence the judgment of a weight on the forehead.
9. Given a sufficiently great difference in thermal conductivity and specific heat, liquids of the same absolute temperature can be discriminated from each other by their temperature impression, just like solid objects. Petroleum, for example, makes a definitely warmer impression than water, whose thermal conductivity is almost four times as great and specific heat approximately twice as high.
10. Ebbecke has called attention to the interesting linguistic parallels, "it is warm, cold" and "it is bright, dark."

Editor's Notes on Chapter I, Division III: The Role of the Vibration Sense in Touch (Sections 39 to 45)

In the next chapter, Katz amasses evidence that the vibration sense is independent of, and superior to, the pressure sense. This is the main point of the volume, and it was discussed extensively above in the Editor's Introduction. There is indeed an independent vibration sense, based on the Pacinian corpuscle, for high-frequency vibrotactile stimulation (100-300 Hz), as well as a system that detects "flutter" (stimulation under 40 Hz) (Talbot et al., 1968). The latter system (flutter), based on cutaneous receptors, might be related to the pressure sense, as von Frey held was true for vibration sensitivity in general. Section 45 is the key section of the present chapter; it summarizes and evaluates the evidence on the vibration sense presented in this and previous chapters.

A few points not made in the Editor's Introduction will be covered here. Katz makes a major distinction between the static-passive senses of color and pressure, and the dynamic-active senses of hearing and vibration (Section 42). He further proposes that color and pressure depend on chemical receptive processes, which result in rapid adaptation and negative after-images, whereas hearing and vibration depend on mechanical receptive processes, which result in minimal adaptation and after-images (Section 41). The latter distinction seems spurious. Although the receptor for the pressure sense has not yet been identified, there is no reason to expect that its mode of reaction, like that of the cone in the retina of the eye, is a complex chemical one. At some level, of course, all receptors respond in a chemical manner.

Katz also distinguishes the pressure sense, as a proximal sense, from the vibration sense, which is in part a remote sense (Section 43). This may largely be true, but it ignores the fact that volume touch (Section 7), which depends on pressure sensitivity, involves remote perception. In fact, the only true proximal percepts are those obtained with the subjective pole, in which the stimulus is not projected "out there."

Katz makes an uncharacteristic concession to the pressure sense (Section 45), when he states that it, not the vibration sense, is primarily responsible for our perception of the tactual surface per se. Pressure provides the basic tactual surface, whereas vibration provides its main properties. Katz repeats the distinction made earlier in the volume (Section 38) between the temporary states and the permanent properties of objects. He devotes a slim section (Section 43) to the perception of vibratory states, and a much larger section (Section 45) to the perception of object properties through the vibration sense.

In the Addendum to Section 43, Katz makes the interesting point that echolocation in bats and the human blind may depend on vibratory sensitivity as well as the auditory sense organ. However, more recent research indicates that the blind perceive obstacles on the basis of auditory information alone (Worchel & Dallenbach, 1947). Katz speculated that the bat's ear is sensitive to both vibrotactile and auditory stimulation, and thus is simultaneously "both an auditory organ and a vibratory organ for inaudible air waves." Likewise, William James (1890) proposed that stimulation of the cutaneous surfaces of the external ear enables the human blind to detect obstacles. To test James' conjecture, Worchel and Dallenbach eliminated auditory sensitivity, but not cutaneous sensitivity, in the ear, by testing deafblind subjects, whose ears did not need to be plugged. They found, contrary to James, that deaf-blind subjects do not possess the "obstacle sense," and are incapable of learning it.

Katz also may have overstated the importance of vibrations in teaching the deaf to speak and understand speech (Section 39). Modern techniques stress the multidimensional nature of the cues present during speech, e.g., changing air pressure from mouth, changing three-dimensional jaw structure, etc. (S. Lederman, personal communication, June 1988).

DIVISION III:
FURTHER ANALYSIS OF
THE TACTUAL PERFORMANCE

Chapter I:
The Role of the Vibration Sense in Touch

Section 39. Vibration Sensations in Neurological
Research and in the Lives of Deaf-Mutes

1. *Neurological research.* Whereas Division I of this report had a phenomenological orientation, and Division II tried to establish experimentally the necessary prerequisites for the tactual accomplishments involving the senses of pressure and temperature, this division considers the extent to which other senses also participate in the fashioning of the world of touch.

As already mentioned, in all tests where sounds produced by touching could possibly influence the judgment on the tactual impressions, the ears were well stopped. Undoubtedly, these touching noises, as well as other acoustical impressions, usually help us develop our image of the touchable world. We excluded the acoustical aspect from our experiments as a source of "error," but, as already mentioned, many observers said that they doubted that auditory impressions had been excluded in spite of our strict measures. It turned out that there was more to such statements than it seemed at first. In addition to the pressure sensations that arise in touching with a moving touch organ, I believe I have established that *there almost invariably are accompanying vibration sensations that also have a role in the tactual accomplishment.* These vibration sensations are closely related in many respects to acoustical sensations, so that occasionally the two types of sensations may indeed be confused. I believe that I can show that the vibration sensations play a very significant role in accounting for numerous results that we have obtained. Above and beyond the domain of these experiments, however, they also play a hitherto scarcely suspected role in the recognition of states of the palpable world. This circumstance motivates the detailed treatment devoted below to vibration sensations, which can be opened here with a few historical remarks on the position of vibration *feelings* (as they were usually called) in earlier neurological research and in the lives of deaf-mutes. The presentation closely follows that of my earlier reports (Katz, 1923a, 1923b), from which much material has been excerpted verbatim.

In his *Medizinischen Terminologie* (Medical terminology) (1919 edition), W. Guttmann defines "vibration feeling" as the ability to feel tuning-fork vibrations on the skin or in the bones. He said that the name (synonym: pallesthesia, from [the Greek] *pallo*, to vibrate) was introduced by Treitel in

1897, whereas the phenomenon itself was first described by Rumpf in 1889. The last statement is not true, however, because von Wittich had already published experiments with tuning forks on the vibration feeling in *Pfluegers Archiv* in 1869. Following up on an even earlier experiment by Valentin, he produced vibration sensations by using rotating disks, whose regularly arranged, raised portions stimulated the skin, and by using struck monochords and resounding reed pipes. Also with regard to Valentin's experiment, Vierordt[1] gave an unfortunately little-noticed warning about using the duration estimated for these individual impressions to determine the duration of the briefest tactual sensation. In 1914, M. von Frey catalogued in Tigerstedt's *Handbuch der physiologischen Methodik* (Handbook of physiological methodology), the procedures that have been used to excite vibration sensations. Besides those already mentioned, he cites periodic contact with a feather, brush, or gear, very weak alternating current, moving rods wound with wire of varying fineness.

A vast number of reports have been devoted to the neurological aspect of vibration sensations. Considerable diagnostic value has been attributed to tests of vibration sensitivity, whose absence may be interpreted quite differently depending upon one's view of its source and nature.[2] In a review whose logical structure is not entirely satisfactory, the Italian researcher, C. Frank,[3] gives some idea of the variety of notions proposed. According to the review, the vibration sensation may represent: 1) a special form of sensitivity, completely independent of all others (Rydel and Seiffer, Bing, Sterling, Mingazzini, Cerulli); 2) not a special sensitivity, but one form of tactual sensitivity (intermittent, tactile stimuli) (Goldscheider, Minor, and Kramer); 3) a modality of deep sensitivity (Marinesco); 4) a type of deep sensitivity that belongs with the other types of deep sensitivity (baryesthesia and bathyesthesia [Egger]); 5) bathyesthesia (Déjerine and Oppenheim); 6) (tactual) surface sensitivity (Treitel, Herzog, Ballieu, Forli, and Barrovecchio); 7) surface sensitivity (tactual, temperature, and pain sensation [Redlich]). Frank himself concluded, based on extensive clinical evidence, that the vibration sensation represents a special form of deep sensitivity, independent of all other forms, and that the nerve fibers designated to carry away the oscillatory excitation probably run in the so-called "motor cable" rather than in the sensory portion of the peripheral nerve trunk.

Frank's literature review is far from complete; it lacks, for example, M. von Frey's important 1914 study on vibration sensations, which, like all of his work, is exemplary in its methodology.[4] Von Frey disconfirmed the view of Egger (*Journal de physiol. et de pathol. gén.*, 1899) and Bing (*Korr.-Blatt Schweiz. Aerzte*, 1910, No. 1) that the vibration sensation depends on the nerves of the bones and not those of the skin. Von Frey believed that he had obtained evidence "that the vibration feeling represents a special manifestation of the *pressure sense* of the skin—one that depends on the type of stimulus, and involves only the nervous structures of this sense." "The examination of the vibration feeling is thus an examination of the pressure sense, which nonetheless may acquire a separate meaning corresponding to the singularity of the stimulus." Mechanical or electrical stimulation produce only intermittent excitation at pressure points; temperature and pain points, on the other hand, respond with lasting, gradually subsiding excitation. In recent publications, to which we will soon return, von Frey has buttressed these findings in a theoretically significant way.

2. *Vibration sensations in the education and the everyday activities of the deaf.* In neurology, vibration sensations lead only an artificial existence so to speak, since, for the present purposes, they must first be produced by rather laborious experimental procedures. On the other hand, they have long been accepted, in a natural form, in the education of deaf-mutes. The teacher has the deaf-mute pupil place one hand on the larynx of a speaking or singing person, so as to obtain a model for their vocalizations, and place the other hand on his or her own larynx. When the same pattern of vibration is received with both hands, then the pupil has succeeded in imitating the sound.[5] In addition, many deaf-mutes come to *understand* speech through vibration sensations without express instruction. A teacher of deaf-mutes, Pfingsten (Kiel; 1802), tells of a deaf-mute who could understand in the dark what a normal girl was saying by placing her hand on the other girl's chest. The most remarkable feat of this type is perhaps that described by J.T. Williams (op. cit.). Willetta Huggins, who has already been mentioned above, could understand the spoken word not only when she placed her fingertips (her middle finger was particularly acute) on the larynx, thorax, or head of the speaker, but even when the vibrations from speaking were transmitted to her fingers via a billiard cue-stick or a sheet of paper. Vibrations also can inform deaf-mutes about nonverbal events involving the objects of their surroundings. In crucial experiments, Eschke[6] showed that electric current, for which true miracles in the treatment of deaf-mutes had been anticipated, has not the slightest therapeutic value. He also provided much data on the sensitivity of absolutely deaf persons to the vibrations they encounter (perception of the striking of a tower clock, a clattering mill, moving wagons, thunder, etc.). As her autobiography indicates, a highly intelligent, blind deaf-mute like Helen Keller may spontaneously make an astonishingly extensive use of vibration sensations in her orientation to the world.[7]

3. *The Sutermeister case.* Helen Keller tells us that she also was able to enjoy music. She said she picked up the melody played on a piano by placing her hand on the instrument. During a visit, W. Stern examined and carried out a few tests on Helen Keller's way of enjoying music, and he concluded that

she really had to experience music in a normal fashion. In all likelihood, her experience was based on the tactually-perceived rhythm and the vibration sensations, which occur in many guises and by their combinations, alternations, and recurrences are capable of providing a sort of aesthetic enjoyment.[8] Recently, the newspapers published a letter from Helen Keller to a radio station in New York in which she described the musical experience provided by the radio, and from which an excerpt is presented here.[9] Even if there is no reason to suspect a great self-deception with respect to her musical sensation,[10] it still may be called a happy coincidence that a new case of musical sensation in the deaf has become known, which is of particularly great scientific significance, considering all of the circumstances.

This deaf person is Herr Eugen Sutermeister, general secretary of the Swiss Society for the Welfare of Deaf-Mutes and editor of the Swiss newspaper for deaf-mutes in Berne. Sutermeister comes from a highly-educated family; at the age of 4 years he completely lost his hearing as the result of meningitis, and soon afterwards his speech as well. Only 55 years after this loss did he discover one day that he could enjoy music; ever since, he has been a music *lover*. He is deeply affected by music and misses no opportunity to obtain good musical enjoyment. I visited Sutermeister in Berne and personally convinced myself of the uncommonly strong effect music has upon him. Together with Professor Révész, I also carried out tests on Sutermeister using organ and piano, and we will report the results elsewhere later. Only the following will be reported for now.

The musical experience probably is based on vibration sensations located in or arising from the chest. The vibration sensations in the hand or in the foot, produced by the resonating floor, have, in contrast to Helen Keller's case, no musical effect, and not only that, they disturb the enjoyment of music. The sensations Sutermeister receives when he puts his hand on a piano are far too strong for him and are not at all pleasing. The excitation in a finger placed on a vibrating cello string is simply awful ("as if I were scratching a piece of slate with my fingernail"). Rhythm plays an important part in Sutermeister's musical enjoyment. His strong rhythmic sense is evident from the highly rhythmic character of poems he has published. Rhythm, however, does not explain everything. Its effect varies, depending upon the type of acoustical material used to produce it. This individual has an unusually strong musical aptitude, so that corresponding findings cannot be expected in deaf-mutes lacking such an aptitude. What has been said here should suffice, so let Sutermeister himself have a brief word:

> A year ago, I was sitting with my wife, who can hear and is a music lover, in the pumproom in Berne, directly in front of an Italian orchestra. All at once, I felt the sound waves streaming toward me in all their chords. I felt as if I had been transported into heaven, and I returned home literally intoxicated by sound.

> As the tones flow through me,
> Clear stream, full of goodness,
> They bathe my soul clean,
> And fill me full of grace. . . .

Since then, orchestral concerts have been one of my greatest pleasures; I already have my favorites among the composers, and many a concert program seems more precious to me than the most opulent menu.

The main receiving station is in my back.[11] Here the sounds penetrate and stream through my entire torso; it is as if I were a hollow metal vessel that was being beaten upon rhythmically and that resounded more loudly or more softly, depending upon the intensity of the sounds. Meanwhile, neither my head nor my hands nor my feet feel the slightest trace; my head is the most unfeeling of all.

I also am good at discriminating the character of the music, whether it is substantial or superficial, joyful or melancholy, solemn or thrilling, monotonous or colorful. My eyes considerably assist my musical sensation; watching the movements of the conductor and the performers, especially the pianists, reveals more easily and rapidly the manner and mode of the music, and better prepares me for what is to come.

The orchestra has to play on a platform. Otherwise, I sense the sounds badly or in a disagreeable manner. They travel along the floor on which both the musicians and I sit, and enter my body through my feet, which is distasteful to me and disturbs my inner harmony.

I first perceived human voices during a performance of the oratorios from Handel's Messiah. There were about 200 singing, and I sensed sounds that were so gentle, so fine, so—how shall I put it?—lyrical, such as I had never heard from a musical instrument—so soft and yet so bright, that I was moved to tears."

4. *Gutzmann's research.* H. Gutzmann[12] deserves special commendation for his systematic investigations on the use of vibration sensations in the instruction of deaf-mutes and others afflicted with speech disorders. He demonstrated the sensitivity of the vibration sense. His subjects, who were normal, but were prevented from hearing, were able to recognize with their touching fingers whole-tone differences in the range A to E on the scale. This precision is important for my hypothesis on the relationship between the senses of hearing and vibration, but it also shows that tone differences are recognized by the touching fingers, which might not have been expected for non-musical persons. However, deaf-mutes, who utilize vibration sensations completely differently than normal subjects, can learn to reproduce emitted sounds very accurately by touching the vibrating larynx; this indicates that the vibration sense is not so very far behind the auditory sense in sensory-motor terms. According to Gutzmann, even totally deaf people who are taught as described, home in *by necessity* to the correct pitch

during innervation of their vocal cords; success does not depend, as one might imagine, on a trial-and-error process that leads more or less haphazardly to the goal. If this observation of Gutzmann's is confirmed, then the vibration sense would provide a highly remarkable parallel to the mechanism hypothesized by Koehler for the direct adjustment of the vocal-cord vibrations to the frequency of the excitation in the cochlea, which we hear as sound.[13] The auditory and vibratory senses also would seem to be related to each other with respect to the structure of the special mechanism that makes possible the sensory-to-motor conversion.

The paths opened up by Gutzmann on the practice of vocal training were followed by R. Lindner[14] and J. Feldt,[15] as well as a group of workers trained by teachers of deaf-mutes in Hamburg,[16] when they sought to improve practical instruction by technical means. The considerable capability of the vibratory sense, as shown in our own studies, makes it easier to use it in a way now to be described to elucidate our own tactual investigation.

Footnotes

1. K. Vierordt, *Der Zeitsinn nach Versuchen* (Experiments on the sense of time). Tuebingen, 1868.
2. Schwaner, under the supervision of Rumpf, appears to have been the first to investigate the use of the vibration sense as a neurological diagnostic procedure. *Die Pruefung der Hautsensibilitaet vermittels Stimmgabeln bei Gesunden und Kranken* (Tuning-fork tests of skin sensitivity in the sick and well). Dissertation, Marburg, 1890.
3. C. Frank, Die Stoerungen der Vibrationsgefuehle bei den traumatischen Verletzungen der peripheren Nervenstaemme (Disruption of vibration feeling from traumatic wounds of peripheral nerve bundles). *Arch. f. Psychiatrie u. Nervenkrankh.*, 62, 1921, p. 712.
4. M. von Frey, Versuche ueber das Vibrationsgefuehl (Experiments on the vibration feeling), *Zeitschr. f. Biol.*, 65. Other pertinent research which Frank did not look into is Head's. A good orientation on Head's research may be found in K. Scholl, Heads Sensibilitaetslehre und ihre Kritiker (Head's theory of sensitivity and its critics), *Zeitschr f. d. ges. Neurol. u. Psychiatrie*, 2, 1910.
5. J. C. Amman clearly expressed the basic principles on the use of vibration sensations in the instruction of deaf-mutes. *Dissertatio de loquela* (Dissertation on speaking), Amsterdam, 1700. The principles were further developed by Georg Raphel (*Kunst, Taube u. Stumme reden zu lehren* [The art of teaching deaf-mutes to speak], Lueneburg, 1718), J. R. Pereire (1715-1780), and Deschamps (Cours élémentaire d' éducation des sourds et muets [Elementary course of instruction for deaf-mutes], Paris, 1779). The interested reader will find more details on this matter in the *Blaettern fuer Taubstummenbildung* (Notes on the education of deaf-mutes), Leipzig, edited by P. Schumann. The following excerpt from Rousseau's *Émile* (Translated by A. Bloom, Basic Books, New York, 1979), is so true to type of the pedagogical intuition present therein that we may allow it to

speak for itself: "In placing a hand on the body of a cello, one can, without the aid of eyes or ears, distinguish solely by the way the wood vibrates and quivers whether the sound it produces is low or high, whether it comes from the A string or the C string. Let the senses be trained in these differences. I have no doubt that with time one could become sensitive enough to be able to hear an entire air with the fingers. And if this is the case, it is clear that one could easily speak to the deaf with music, for sounds and rhythms, no less susceptible of regular combinations than articulations and voices, can similarly be taken for the elements of speech" (p. 138).

6. E. A. Eschke, *Galvanische Versuche* (Galvanic experiments), Berlin, 1803.
7. H. Keller, *The Story of My Life*, New York, 1905, Grosset & Dunlap. "I used to make noises, keeping one hand on my throat while the other hand felt the movements of my lips. I was pleased with anything that made a noise and liked to feel the cat purr and the dog bark. I also liked to keep my hand on a singer's throat, or on a piano when it was being played" (p. 58). "They forget that my whole body is alive to the conditions about me. The rumble and roar of the city smite the nerves of my face, and I feel the ceaseless tramp of an unseen multitude, and the dissonant tumult frets my spirit. The grinding of heavy wagons on hard pavements and the monotonous clangour of machinery are all the more torturing to the nerves if one's attention is not diverted by the panorama that is always present in the noisy streets to people who can see" (pp. 123-124).
8. W. Stern, Helen Keller, *Zeitschr. f. angew. Psychol.*, *3*, 1910, p. 327.
9. "Last evening, as my family was listening to the immortal work in my home, I laid my hand on the receiver—and clearly felt the vibrations. Then I had the cover unscrewed and lightly touched the diaphragm. How amazed I was to discover that I could feel not only every vibration, but also the passion of the rhythm, the pulsation and swelling of the music. The fusion of the vibrations from the different instruments enchanted me. I could precisely distinguish between the cornet and the rolling of the drums, the deep tone of the cello and the singing of the violins. How lovely did the song of the violins flow over the deep tones of the other instruments! And when the human voice broke through the surging harmony, I recognized it immediately. I heard the chorus swell jubilantly, become more and more ecstatic, and blaze high like a flame— and my heart stopped. The women's voices seemed like an incarnate choir of angels as they flowed on in a harmonious flood of the purest beauty. And the ebb and flow of the great chorus beat sharply against my fingers. Then instruments and voices joined together—an ocean of wild vibrations—and died away like the breath of a mouth trembling in sweet softness."
10. As W. Stern points out in his review on the psychology of sensorially-deprived persons (*Zeitschr. f. angew. Psychol.*, *3*, 1910, p. 562), W. Jerusalem, with Laura Bridgman, and William Wade, with three other deaf-blind American girls, have verified the pleasure obtained in the playing of a hand-held music box.
11. This statement surely reflects a misinterpretation of the experience, which depends on the resonating thorax.
12. H. Gutzmann. 1. Ueber die Bedeutung des Vibrationsgefuehls fuer die Stimmbildung Taubstummer und Schwerhoeriger (Significance of the vibration feeling for vocal development in deaf-mutes and the hearing-impaired), *Ver-*

handl. der deutschen otolog. Ges., at the 15th meeting in Vienna, 1906. 2. Ueber die Grundlagen der Behandlung von Stimmstoerungen mit harmonischer Vibration (Basis of treatment of vocal disorders with harmonic vibration), *Archiv f. Laryngol. u. Rhinol., 31*. 3. Untersuchungen ueber die Grenzen der sprachlichen Perzeption (Studies on the limits of speech perception), *Zeitschr. f. klin. Medizin, 60*, 1907.

13. W. Koehler, Psychologische Beitraege zur Phonetik (Psychological contributions to phonetics), *Katzensteins Archiv f. exper. u. klin. Phonetik, 1*, 1913. From my own acoustical experience, Koehler's hypothesis seems very plausible. Stumpf also largely agrees with him. *Zeitschr. f. Psychol., 94*, 1924, p. 7. O. Laubi (Der Rhythmus und seine therapeutische Verwendung [Rhythm and its therapeutic use], *Zentralblatt f. d. ges. Neurol. u. Psychiatrie, 82*, 1923), states that when the notes sung by the phonasthenic patient do not have the same frequency as the rhythm applied therapeutically at the sternum, *beats* arise that are sensed as jolts and serve to correct the innervation. This would seem to answer affirmatively the question of whether anything like beats occurred in the vibration sense, which Mr. E. von Hornbostel directed to me during an exchange of letters on the vibration sense. An investigation by K. Hansen and P. Hoffmann on the chain of reflexes produced by vibration in normal and ill subjects (*Zeitschrift f. Biol., 74*, 1922) also throws some light on the triggering of the sensory-motor mechanism in the vibration sense. They showed that when the hand or the foot of a normal subject is placed on a vibrating wooden rod, then in the muscles rhythmically stretched by the rod, as indicated by the electromyogram, a synchronous chain of reflexes is produced only if the subject simultaneously innervates the muscle voluntarily and thereby facilitates the reflex. Very likely, the deaf-mute also facilitates the reflex produced at the vocal cords by means of a voluntary, but not precisely tuned, innervation. In voluntary tetanus, the number of vibratory beats is 39 to 48 per second, according to recent work by A. V. Hill (*Journal of Physiol., 55*, 1921).

14. *Der erste Sprachunterricht Taubstummer auf Grund statistischer experimenteller und psychologischer Untersuchungen* (Initial instruction in vocalization for deaf-mutes on the basis of statistical, experimental, and psychological studies), Veroeff. d. Inst. f. exp. Paed. u. Psych. des Leipziger Lehrervereins, 1910, 1.

15. Der Tastreifen, ein Hilfsmittel zur Verbesserung des Sprechens der Taubstummen (The tactual flute, an aid for improving the speech of deaf-mutes), *Blaetter f. Taubstummenbildung*, 1913.

16. A. Schaer, *Ueber den Tastsinn und seine Beziehungen zur Lautsprache* (The sense of touch and its connection to speaking aloud), Nos. 1-2, Vox, 1922.

Section 40. Vibratory Thresholds: Rejection of Visual Analogies

In deaf-mutes, the vibration sensation can substitute fully for the missing auditory sensation. In hearing persons, in many cases at least, there is not just the actual external, even if very intimate, coupling of acoustical and tactual elements, but rather a deep, phenomenological and structural *relationship* of the acoustical and the vibratory sensations. The evidence for this will be presented in detail below. As early as 1846, Weber (op. cit., p. 118) pointed out the transitional experiences from touch to hearing: "When the beats occurring in rapid succession on a touch organ fuse into one sensation, but the interval between beats affects the sensation, we have a transition from touch to audition. We feel the beating as a trembling, which we perceive with the ear as a sound. This trembling is capable of the most varied nuances, which one senses very well when ice skating, in which different variations in the sensation are perceived from the smoothest to the roughest ice." Weber's words should have received more attention after the introduction of Darwin's evolutionary viewpoint led to a search for transitional forms of the acknowledged human senses. I myself now see the vibration sense, with its special sensory modality, as a stage in the development leading from the sense of touch, as this word is currently understood, to the sense of hearing.[1] From this viewpoint, the vibration sensation also loses the character of a curiosity, which hitherto has often been attributed to it.

Tests of the tactual sense with tuning forks or other vibrating devices were indeed all dominated by the notion that the phenomena occurring in these experiments, by analogy to visual fusion phenomena, could provide no other information than that on the fusion of brief tactile sensations; this notion has blocked insight into their true nature. If the vibrations of a tuning fork with a moderate frequency are applied to a pressure point of the skin through a stiff bristle, and one does not take care that as far as possible no oscillations occur in the skin, then not only will the pressure sense be excited, but the vibration sense as well, and vibration sensations will accompany the pressure sensations. In one such experiment, the analogy to visual fusion held only for the pressure sense; the fusion of the successive pressure stimuli actually occurred rather rapidly. Mach[2] once stated that the interstimulus interval at which this occurs is .0277 sec; the corresponding values are .0470 sec for the eye and .0160 sec for the ear. "During mechanical stimulation of the tip of the index finger with successive blows, . . . the interval, i.e., the time between the onset of the first stimulus and onset of the second, would have to be at least about .05 sec if the two beats are to be recognized as separate."[3] Note well, however, that in such experiments there is only *one* threshold, that for fusion. Thus, just as in the visual case, a further shortening of the interstimulus interval after fusion has been achieved no longer changes the impression for tactual stimuli that excite only the pressure sense. It is completely different for the vibratory sense. If a tuning fork with a low frequency produces vibration sensations in the fingers, then their character changes, as can be seen from Gutzmann's experiments cited above, when one switches to a tuning fork with a substantially higher frequency. Here the proper analogy is not to vision, but to audition.[4] Two points stand out in the stimulus series. Just as the effective oscillations in audition are bounded by two thresholds—below the lower threshold the vibrations are not yet perceived as tones and above the upper threshold they are no longer perceived as tones—so, too, the effective oscillations for the vibration sensations lie between two thresholds. The precise thresholds have yet to be determined. My preliminary findings indicate that for the finger, the lower threshold is below 50 Hz, and the upper above 500 Hz. Other parts of the body have different thresholds. Almost all of the extensive data found in the literature refer to the *upper* threshold.[5] The thresholds I have spoken about here mean something quite different than threshold in the sense of "minimum perceptible." Just as one can ask what the lowest sound intensity is that just barely excites the ear, so, too, it makes good sense to determine the necessary energy for the "minimum perceptible" of the vibration sense.[6] I have not made more precise measurements, but I rather think that the minimum vibratory threshold (like the auditory threshold, it will depend upon the type and amplitude of the vibrations) for deaf-mutes under the most favorable conditions little exceeds that of the ear for normal subjects.[7] Tapping experiments have shown me that this threshold has extraordinarily different values on different parts of the body. It is well known that the human body is very sensitive to the jolts of earthquakes, and a significant component of this sensitivity is vibratory in nature. Determining the differential vibratory threshold is just as important as determining the absolute threshold. Although rigorous research is yet lacking, casual observation nevertheless indicates the number of just noticeable differences to be very considerable.

Footnotes

1. According to Muck, the fact that shock in humans and the cataleptic fit in animals blots out the sense of hearing (and pain), but not that of touch, also indicates that the sense of touch is phylogenetically older than the sense of hearing. O. Muck, *Die seelische Ausschaltung des Gehoer - und Schmerzsinns bei Mensch und Tier als Parallelvorgaenge im Licht der Phylogenie betrachtet* (Psychic disconnection of the senses of hearing and pain in humans and animals as parallel processes phylogenetically). *Muench. med. Wochenschr.*, *67*, 1920. Wundt emphatically advocated the principle of adaptation of the sensory functions to the stimuli, and of the sensory mechanisms to the functions. This chapter confirms the principle for a special case. M. Ettlinger has energetically employed the phylogenetic viewpoint in very noteworthy expositions on the theory of specific nerve energy. *Bericht ueber den 5. Kongress f. exper. Psychologie* (Report on the Fifth Congress for Experimental Psychology). Leipzig, 1912, p. 235 f.
2. E. Mach, *Sitzungsber. der Wiener Akad., math-naturwiss. Kl.*, 2 Abt. (Section 2), *51*, 1865.

3. A. Basler, Ueber die Verschmelzung zweier nacheinander erfolgenden Tastreize (Fusion of two successive tactual stimuli), *Pfluegers Archiv, 143,* 1911, p. 244. In the same volume of *Pfluegers Archiv,* Basler expressed his belief that touch is comparable to vision because for both senses there is more rapid fusion of double stimuli than of longer series. Nevertheless, aside from our objections above, the large disparity between vision and touch in the ratio of fusion times for double and serial stimuli appears to rule out drawing such a parallel. The ratio for vision, according to Basler, is about 1:3, whereas that for touch, calculated for many positions of the skin, is less than 1:30.
4. As for the summation of intensities, the question remains open on whether there is a tactual analog to Talbot's law in vision.
5. On this, see, e.g., Schwaner's (op. cit.) numerical data. According to von Frey in Tigerstedt's *Methodik,* V. Grandis found that when the skin was stimulated by a very weak alternating current, the initially discontinuous sensation changed after awhile into a continuous touch sensation.
6. J. E. Wood determined the minimum vibratory threshold in healthy and sick subjects, and found that even at this initial stage of processing the threshold at the sacrum and the spina iliaca ant. sup. was higher in 80 tabetic patients than in 100 normal subjects. *American Journal of the Medical Sciences, 163,* 1922.
7. Therefore, the minimum vibratory threshold presumably would require less energy than the minimum pressure threshold. Studies by von Frey indicate their magnitude, *Zeitschr. f. Biol.,* 70.

According to Ewald's studies on the end organs of the VIII (acoustic) cranial nerve (Wiesbaden; 1892), even with bilateral loss of the labyrinth, deaf persons still respond to many auditory stimuli; Hermann and Bernstein interpret this to mean that animals still perceive the oscillation of the air through the skin (that is, by means of vibration). Even if Ewald himself had considered such a possibility and had attempted to preclude it by appropriate experimental procedures, it still could not be regarded as precluded, it seems to me, given the astonishingly high sensitivity of the human skin for vibrations.

Section 41. The Independence of Vibration and Pressure Sensations

An electromagnetic tuning fork with a moderate frequency serves best for an initial probe of the laws of the vibratory sense. If the tuning fork is fastened on an iron stand, then the impulses from its vibrating free end can easily be collected. Portions of the arm, leg, head, or torso, when brought into contact with the stand, all respond more or less clearly with vibration sensations. In doing this, one confirms von Frey's finding that the vibration sensations are clearer at points with tight, stretched skin than those with soft, relaxed skin, and clearer at points with bone close beneath the skin. I agree with von Frey that this is due to the greater resonance capacity of bones and stretched skin. Because of bone conduction, the vibration sensations often occur both at the point of stimulation and at distant points, and often occur even more strongly at the distant points. The remarkably weak vibration sensation on the tongue no doubt is due to its low resonance capacity.

The apparatus just described is also suitable for isolating vibration sensations and thus demonstrating their independence from pressure (contact) sensations. If the elbow is rested on the stand, then the vibrating tuning fork produces vibration sensations in the hand (via bone conduction) in the complete absence of pressure sensations there. This and the following experiment also show that it would be incorrect to regard the vibration sensations as movement experiences within the pressure sense. When the fingers are placed loosely on the stand, then as long as the tuning fork remains still, a tactual sensation is present. When the tuning fork begins to vibrate, then the vibration sensation is added as a completely new and isolated impulse localized in the fingers. By no means does one sense that the stand is moving or that the pressure sensation present waxes and wanes periodically.[1]

Yet other results indicate a sharp separation of pressure and vibration sensations. If the stand holding the tuning fork is touched, then the vibration sensation clearly occurs later than the pressure sensation. If the contact is only brief, say .5 sec, then no vibration sensation at all occurs, whereas it indubitably occurs with longer contact. Either the sensory organs themselves have a corresponding latency period vis-á-vis the oscillations provided here, or a certain length of time is needed to overcome resistance in more central segments of the nervous system. This says nothing about the latency period for other types of vibrations than those of the tuning fork, that is, for example, those to be mentioned below (Section 42).

The following experiment also supports the independence of the vibration sensations from the pressure sensations. With the tuning fork vibrating weakly, place your fingers quite loosely on the stand, so that the pressure exerted definitely does not exceed 5 g. The vibrations then will be weakly, but clearly, felt. Now increase the pressure on the stand, say to about 5,000 g. Although the pressure has now increased a thousand-fold, the vibration sensations nevertheless remain approximately equally clear; even *during* the increase in pressure, the character and intensity of the vibration sensations do not change in any clearly discernible fashion. The increasing pressure of the hand somewhat decreases the amplitude of the vibration and thereby the intensity of the vibratory excitation.[2] Our weak vibration sensation decreased as the pressure increased, but if it was nothing but a moving pressure sensation, then its persistence in the background of a greatly increased pressure sensation could not be understood at all.[3] Here again we see that the analogy to vision breaks down: if a light is weakly flickered on a colored surface that has its own independent source of illumination, then it usually takes little additional illumination to weaken the flickering and finally make it completely invisible. If, as von Frey claimed, the vibration sensations are tied to the organs of the pressure sense, a view to which I, too, am inclined, then our last experiment indicates that these organs can take up two states of excitation absolutely simultaneously and thus their readiness to transmit vibration sensations is independent of the state of excitation resulting from an increasing or a constant pressure stimulus.[4]

Based on views developed below (Section 45), I will take as given the double-excitation state of the pressure sense organ during each touching of roughness that is accompanied by a vibration sensation. In the numerous instances where an excitation state of that sort provides us simultaneously with information on permanent properties of the tactual objects (hardness, softness) as well as on transitory appertinents (vibrating), the pressure sense organ reveals a hitherto unsuspected *analytic* capability.

For the pressure sense, like the eye, rapid fatigue and after-images occur in the presence of motionless stimuli. The two senses correspond in this regard because in both cases chemical processes are crucial for the functioning of the peripheral sensory apparatus. The much lower fatigue and the complete absence of negative after-images in the ear is explained by its reliance on mechanical processes; in the final analysis, it is the vibrations of the basilar membrane that determine the operation of the peripheral sensory apparatus. If the vibratory sense exhibits neither decided fatigue (you can rest your fingers for ten minutes or more on the vibrating stand without being able to discern a definite loss in clarity of the vibration), nor definite after-images (so far as I know, only after stimulation of the lips with a tuning fork does a rapidly subsiding, tickling-vibrating after-sensation persist[5]), then this indicates that, as with the ear, mechanical processes in the skin exercise a decisive influence on the sensory organs. Through energetic rubbing, such as on a band of beads, one can temporarily increase the pressure threshold of a finger enormously, but seemingly impair vibration sensitivity much less thereby.

Footnotes

1. In this experiment, the eye can infer from the blurring of the contours of the tines that the tuning fork is moving. However, in the stand itself, with its minimal movements, that cannot be seen in any way. Here again, therefore, the touching hand is shown to surpass the eye. It may be mentioned, too, that the smallest movements determined by G. Stoerring (*Archiv f. d. ges. Psychol.*, 25, 1912, p. 178), which are those felt in the elbow joint (von Frey, *Ber. d. bayr. Akad. d. Wiss., math.-phys. Kl.*, 1918, p. 95, calculates them to be .025 mm at a distance of 29 cm from the axis of rotation), are imperceptible to the eye.
2. The significance of the amplitude of vibration for the vibration sensation is discussed by G. Sergi, *Zeitschr. f. Psychol.*, 3, 1892.

3. We are reminded here that the soles of the feet can perceive vibrations in the floor, even though the entire weight of the body is pressing on them. I have not succeeded as yet in eliminating the objection that here it is not the sensory organs subjected to the maximum pressure, but rather those at some distance, which are pressed more weakly or not at all, that respond, via resonance, to the vibration. This does not sound very plausible, but I wanted to mention it nevertheless. The objection could, of course, also be raised against the experiment involving an increase in pressure at the fingers.

   According to a view developed by F. Kiesow, the deformation of the skin under pressure produces a pressure gradient in its fluid and thereby a change in its chemical composition, which represents the actual chemical stimulus for the sensory organs of the pressure sense. Thus, the sensitivity of the sensory organs to vibrations would also be independent of these chemical stimuli.

4. This is not to say that the converse also holds, that is, that pressure sensitivity is independent to the same extent of the state of the sensory organs induced by vibratory stimulation. The effect of vibratory massage (relief from chronic, aching excitation with vibrators or vibrating arm-chairs) tends to speak against this. On this, see E. Plate, Ueber einen Vibrator mit erhoehter Erschuetterungszahl (A high-frequency vibrator). *Zeitschr. f. physik. u. diaetet. Therapie,* *17,* 1913.

5. The only apparent disproof of this is what Sutermeister once said: "I regret very much that all too often one piece of music follows too quickly after another. What has been played reverberates long and loud in me, so that I even imagine during the pause that the music is still being played. If the new piece of music then arrives, the old and new sounds mingle within me, and a disharmony ensues." We will look more closely at this matter in a prospective study.

Section 42. The Form of Vibratory Stimulation, and the Temporal Pattern of the Vibration Sensation

If M. von Frey regards vibratory excitation, and not prolonged deformation, as the true adequate stimulus for the pressure points,[1] then this certainly holds insofar as the vibration sensation is concerned. Like the auditory sense, the vibration sense has an oscillatory form of stimulation. The auditory organ can be excited by sound waves that reach it through the air, through a liquid (e.g., hearing under water), or through the solid body (e.g., a vibrating tuning fork placed against the upper teeth). The vibration sense can be excited not only by oscillations of solid bodies (see the preceding section), but by oscillations of liquid ones as well. A tuning fork operating in water produces vibration sensations in a finger that is immersed not too far away.[2] What about the excitation of the vibration sense through air waves? Except for cases occasionally cited in which deaf-mutes supposedly notice the sound waves striking the hand, this only occurs in normal subjects when vibration sensations are produced in the chest by certain low organ tones.

Is there a difference between [pure and chaotic] vibration sensations corresponding to that between the auditory sensations of tone and noise? To settle this question, I set up the following experiment. A motor-driven wooden disk spins against one end of a firmly-clamped wooden rod. The friction produces vibrations in the rod, which can then be picked up by a touch organ placed upon its other end. The vibrations could be varied in type and intensity by inserting various sorts of material between the disk and the rod. The variety of noises thus created provides an acoustical control. Vibrations produced in this way can be compared with each other, as well as with those produced by applying a vibrating tuning fork to the wooden rod. The latter comparison, which is the key one here, reveals a clear difference between the vibration sensations produced by friction with the motor and those produced by the tuning fork. One may characterize the latter as finer, more ordered, more gentle, and more pleasant. I believe that I was no victim of autosuggestion when I concluded from the comparison that the difference between the vibration sensations is related to that between noises and musical tones in acoustics. Completely trustworthy results could be obtained here only if strictly blind experiments were carried out with intelligent and psychologically well-trained deaf subjects. I have not been able to obtain such subjects. One more finding from the experiments also should be reported, because it will be important in the interpretation of later observations: the latency of excitation of the touch organ is much greater with tuning-fork vibrations than with those produced by friction.

It is not now possible to resolve definitely the related question as to whether there are vibratory analogs of the aural harmonics; a few observations presented below (Section 43) indicate there are.

Is there an "absolute experience of vibration" corresponding to the absolute experience of tone? This must indeed be the case, since otherwise deaf persons could neither recognize a sound from its vibrations nor properly reproduce a desired sound. Likewise, the development of an absolute memory for vibratory impressions in normal persons is indicated by those cases involving absolute judgment of roughness, which evidence presented in more detail below will show to be based on vibration sensations.

The temporal pattern of appearance characteristic of vibration sensations clearly separates them from pressure sensations, but links them to auditory sensations. The temporal garb in which all experiences of consciousness, including the sensations, appear has, of course, been subjected time and again to clever philosophical scrutiny, but it nevertheless seems as if too abstract an approach has often allowed the considerable differences in temporal pattern between different sensations to be overlooked.[3] We experience an incessant waxing and waning of a perfectly uniform tone, which varies in neither intensity nor quality. A presently experienced fragment of the tone sinks into the past to the same extent that new material rises on the other side. Here we experience in the strict sense of the term a *process*; had it not become so well worn by general use, we would have reserved the expression "stream of consciousness" for the sensing of the temporal structure of the tone sensations. A color experience lacks this temporal structure. The color presented in a motionless visual field does not come and go like the tone. For the color experience, we cannot speak of a process; a color is experienced as something continuously there. (We are thinking here only of the color *phenomenon*; at a physiological level, of course, we naturally assign responding processes to all sensations in the nervous system.) To speak of a process in the case of color, some clearly marked change must occur. However, in an unchanging visual field, color has a completely different character than tones. I maintain that pressure sensations belong with the temporally stationary sensations, as exemplified by color, whereas vibration sensations have exactly the same temporal character as tones. A pressure experience of the motionless finger, like a color, is simply there, without waxing or waning; the sensation of a process is aroused only by a change that is far more dramatic than the waning of sensation that results from adaptation of the motionless touch organ. The statements above on tones hold as well if we substitute "vibration sensations" wherever "tone sensations" occurs. We have already refuted the notion that vibration sensations are merely experiences of movement occurring within the pressure sense. This quite natural mistake can be explained by the statements above. If the pressure sensation changes, then a true process is experienced. Set the fingers on a table top, and if vibration sensations join in with the pressure sensations at the instant when a tuning fork begins to vibrate, then it is natural to associate the waxing and waning of the vibration sensations with the pressure sensations, which actually remain constant.

Tones and vibrations, colors and pressure sensations are not as a rule given to us as neutral phenomenal entities to advance psychological research, but as harbingers of the appertinents and properties of the objects in our environment. If the information takes the forms of appearance of tones and vibrations, then we will speak of a dynamic-active mode,[4] and otherwise of a static-passive one. Later statements will make use of this distinction.

Footnotes

1. *Sitzungsber. d. physik.-med. Ges. Wuerzburg*, 1919, p. 6. Von Frey has pointed out that what appears when an isolated pressure point is touched is equivalent, except for the much briefer duration, to the buzzing sensation produced by electrical stimulation at this point. *Psychologische Forschung*, 3, 1924, p. 212. Goldscheider and P. Hoefer agree with him: "The lightest tactile stimulation of

a pressure point produces an oscillating, *buzzing* sensation, which likewise is discernible from careful superficial stimulation of skin areas abundant in pressure points." Ueber den Drucksinn (On the pressure sense), *Pfluegers Archiv, 199*, 1923, p. 295. On this, see also the statements by von Frey and Kiesow: "The excitation of the tactile nerves is generally tetanic. Only because of this property is it possible to recognize the prolonged deformation of the skin as such." *Zeitschr. f. Psychol., 20*, 1899, p. 156.

2. The experiment also works with mercury. I mention this in regard to Meissner's experiment, in which, as is well known, the hand immersed in mercury does not feel the constant pressure under the surface.

3. Some very useful observations pertaining to this point are provided by H. Lotze, *Medizinische Psychologie* (Medical psychology), Leipzig, 1852, p. 378. "By its very nature, sound is inseparable from the flow of time, but has nothing at all to do with space. On the other hand, the essential nature of color would be apprehensible to us even if it possessed no specifiable temporal duration. . . . If sound is a live event, comprehended while in progress, color is a motionless state." A few statements by H. Plessner, *Die Einheit der Sinne* (Unity of the senses), Bonn, 1923, p. 214, also are pertinent here: "Only a sound swells. Light, on the other hand, has a static character. An increase or decrease in intensity is manifested in audition as a swelling up or down, but in vision as a mere brightening or paling of the phenomenal illuminated surface. While any change in sound intensity has the sensation valence of a change in volume, this characteristic feature is lacking in vision. "Differences in the temporal pattern of the sensory phenomena have also found their way into the language. 'The stone is hard, the meadow is green, but the brook babbles, the thunder peals. . . . Thus, acoustical properties are expressed in the verb." W. Haas, *Die psychische Dingwelt* (The world of mental objects), Bonn, 1921. Exceptions such as "the material molts" and "the heaven darkens" are rare, and we will not go into their roots in greater detail.

4. Vibration sensations possess more energy than many other sensations, which enables them to impress themselves on our consciousness, and thus supplant other sensations. This may explain not only the effect mentioned above (Section 41, Footnote 4), but also the pain relief that has opened the way for the therapeutic use of vibration. We have not considered at all here the therapeutic value attributed to vibratory treatments in circulatory and related disorders.

Section 43. Perception of Vibratory States Through the Vibration Sense

By its nature, the pressure sense is a proximal or near sense, that is, the object must be in direct contact with the skin to be perceived. Although we obtain certain information on the shape, hardness, and other properties of objects not directly in contact with our body when we touch them with a stick or a probe, that is quite certainly based on acquired associations in these and similar cases that are exceptions to the rule. The vibration sense is not a proximal sense in the same sense. When we perceive vibrations on the floor of a room from a tuning fork several meters away, or from a machine at a far greater distance under some circumstances, then here, too, just as in audition, oscillations serve as a mediator between us and the source of stimulation. Any number of individuals can be affected by vibratory stimulation, just as with auditory stimulation. Thus, the vibration sense enables touch to free itself from its confinement, and to begin to conquer distance. Touch thereby shows an interesting *duality*; one branch, the pressure sense, is a proximal sense, whereas the other, the vibration sense, has become a distal or far sense. It is now generally assumed that the different senses have evolved from *one* primitive sense. I propose instead that the vibration sense represents one step in the evolutionary sequence leading from touch to hearing.[1] Animals responded earlier to oscillatory stimuli with vibratory than auditory sensations. It was progress indeed when the animal began, through vibration sensations, to have experience of events lying beyond the boundary of its own body. In this way, it could receive early warning of an approaching enemy or prey. The vibration sense, to be sure, was not able to register precisely the weakest oscillations of the air at a great distance, and the development of a highly refined apparatus, such as the human ear, was necessary. The hypothesis just presented can perhaps help provide some notion of the capability of more primitive auditory organs.[2]

Strictly speaking, a stimulus of a proximal sense can never be made accessible to more than one observer, while in principle any number of individuals can share a stimulus of a distal sense. Only one person can taste the same piece of sugar, or feel the head of the same needle at the same time, but a cannon shot can be heard by millions. The vibration sense, as a distal sense, is likewise a social sense. In war, the same explosion is felt by countless individuals as a vibration on the ground.

Discussed below will be the ability to perceive vibratory events on our own body or on objects in our environment. How the vibration sense is used in speech training for the deaf has already been touched on, but to say that vibration sensations are vital for the training of the sensorily handicapped, is not to deny that they are highly significant for the speech and singing of normal persons as well. The vibrations produced by the oscillations of the vocal cords are clearly perceived subjectively due to the resonance of our body; we localize them in the chest, neck, and head.[3] It seems very likely to me that the vibration sensations so perceived are more significant for the proper innervation of the vocal cords than are the oft-discussed kinesthetic sensations of the larynx.[4] Recognition of this might prove useful in the systematization of speech and vocal training for normal persons as well.

We turn now to the perception of vibrations on objects. Vibration sensations will tell us about the heavy wagon passing in the street, and of the motion of the vehicle in which we sit; they help the captain in handling the moving vessel, the pilot in controlling the airplane. Once attention is paid to this type of phenomenon, one discovers that ours is not only a resounding world, but also, in great measure, a vibrating world.

A train ride is particularly rich in opportunities to study vibration sensations. Suppose the train is traveling on a straight stretch at a uniform speed, so that the vestibular organ comes little into play. We stop up our ears really well, close our eyes, and then sink into the inrushing symphony of vibration sensations. The steady vibratory impressions, some more delicate and others coarser, stream along, but without interfering with each other. Set into relief against this backdrop are the softer vibration waves and the sharper vibration impulses. The sensations gain entry through very different parts of the body, e.g., through the soles of the feet, the thigh, the buttocks, the back, and perhaps even the resting hands and arms. The different streams are merged in consciousness, and even if the unity is less perfect than with the aural harmonics, one can scarcely avoid stating that the vibratory elements are fused by consciousness into more complete units.

The observations stated above on the accomplishments of the vibration sense perhaps have some import for medical practice as well. The neurologist may consider them in the examination of vibratory sensitivity, and the otologist must always allow for the very extensive substitution of vibration sensations for hearing sensations.[5]

Footnotes

1. In this formulation, I wish to allow for the possibility, mentioned by von Hornbostel (*Psychologische Forschung*, 5, 1924, p. 370), that the vibration sense did not differentiate itself from the pressure sense, but rather the opposite occurred, that is, the pressure sense differentiated itself from the vibration sense.
2. "At the lower frequency threshold, tone sensations shade into fluttering and at the upper intensity (and frequency) threshold into tickling (then into pain), that is, into vibration sensations. One is absolutely forced to regard the auditory organ as a more highly evolved sensory organ for vibration (von Hornbostel, ibid.)." "In hearing, the genetic connection with the vibration perception of a statolith organ is even discernible in introspection; at the lowest tones with the slowest vibrations, the ear is still able to perceive the individual sound impulses." M. Ettlinger, *Tierpsychologische Anmerkungen* (Remarks on animal psychology), p. 235 f.
3. Emit some sounds with the hand held on the larynx in order to experience for yourself what has been said. In doing this, compare the vibrations localized inside the body with those perceived on the fingers.
4. Recently, other authors also have repeatedly cautioned against overestimating the value of kinesthetic sensations, e.g., T. Ziehen (*Zeitschr. f. paed. Psychol. u. exp. Paedagogik*, 10, 1914, p. 43); see also Ziehen's valuable survey on the history of the theory of kinesthetic sensations, *Fortschritten der Psychologie* (Advances in psychology), Vol. 1, 1913. —"The internal subjective tactual sensations are

often overestimated, whereas the external ones, perceived on the object, have been underestimated." R. Lindner, *Untersuchungen ueber die Lautsprache und ihre Anwendung auf die Paedagogik* (Studies on speaking aloud and its pedagogical application), Leipzig, 1916, p. 85.

5. A procedure occasionally used by otologists to unmask people who play deaf should be mentioned here. Drop a heavy object behind the back of the true deaf, and as a rule they will turn around, because the vibration impulse through the soles of the feet, which they have learned to attend to, informs them of the fall. Pretenders let the falling go entirely unnoticed because they are not permitted to *hear* it. As hearing persons, they have not learned to attend to the accompanying vibrations at all, or do not know how to fit them into their dissimulation scheme.

Addendum. Vibration Sensations in Animal Psychology

In animal psychology, vibration sensations can help to explain the behavior of animals that clearly react to acoustical stimuli without having auditory organs, or that do have auditory organs, but evidently let their behavior be determined more by mechanical shocks in the surrounding medium than by acoustical stimuli. Can fish hear? Von Frisch recently delved into this age-old question and convincingly demonstrated that fish do react to acoustical stimuli.[1] It now only needs to be established whether the auditory nerve is excited when this occurs or whether the fish really are reacting to vibration sensations. It is well known that lizards take flight when the ground shakes, but not from acoustical stimuli. G. Kafka[2] writes: "The rheotropic-anemotropic phenomena also include cases in which a solid object reflects back to an organism the water or air waves produced by the organism's movement in approaching it. Such touching at a distance, which is well known in the human blind, has been described particularly in the case of insects, fish, and bats." Here, surely, vibration sensations are involved. More recent research indicates that with bats, "a transmission of the air-pressure fluctuations to the eardrum is required" (Kafka). If the bat's ear also is sensitive to tactile-vibratory stimuli, then it would turn out to be, simultaneously, both an auditory organ and a vibratory organ for inaudible air waves.

According to O. Koehler (Sinnesphysiologie der Tiere [Sensory physiology of animals], *Jahresbericht ueber die ges. Physiologie,* 1922, p. 435), Rabaud found a highly-developed vibration sense among different web-weaving spiders, which are attracted by a chambered tuning fork (*Kammer-a-Gabel*). Even if the instrument is placed not at the edge of the web, but 1 cm away, the vibration of the air is plainly sufficient to attract the spider. In a case where the sense of hearing can be regarded as excluded, A. Dichtl has shown that the ant, *camponotus herculaneus,* reacts very precisely to the shaking of the ground. "The vibration sense, which alerts them to even the slightest movement of the ground, certainly plays a very prominent role in the lives of ground-dwelling animals that move by creeping. This sense can signal the approach of danger from a far distance" (*Zeitschr. fuer wiss. Insektenbiol.*, *19*, 1924).

Matthes demonstrated very convincingly the role of the vibration sense in the acquisition of food by salamanders. "The fact that the vibration sense by itself can trigger an alarm reaction may be most simply demonstrated using blinded salamanders, which, by means of a gentle stirring of the water, such as by moving a glass rod rapidly to and fro, are rudely jolted from their rest. However, the vibration sense also can serve as a guidance device. When we produced the vibration to the right of the salamander's midsection, for example, it turned by jerks to the right until it stood with its nose directly in front of the moving rod and snapped at it. However, if we do not let it get quite as far, and move slowly on with the rod, then the salamander follows behind it."[3] Haecker previously obtained similar findings in methodologically less incisive experiments. "In obtaining food, the axolotl [a Mexican salamander] is probably guided by a special function of the oral sense, and very likely by the reception of gentle stirrings of the water."[4] According to the observations made by Matthes, there is a striking *localization* of the center of vibration, a feat that bears comparison with sound localization.[5] At a lecture on the vibration sense, a zoologist informed me of observations which indicate that the behavior of hares and wild rabbits is substantially influenced by vibrations reaching them through the ground on which they sit. Indeed, they even obtain vibratory information as to the direction from which the danger threatens. I could not collect first-hand information on this myself, but the findings do not sound implausible, because among deaf persons an approximate localization of vibration foci likewise occurs under favorable conditions, though, to be sure, it attains nowhere near the perfection of sound localization.[6]

Let us delve briefly into the role of the sense of touch, in the original sense of that term, in animal psychology. The perception of surface structure does not appear to be very important biologically for most animals, at least not for mammals and birds. The hairy or befeathered portion of the animal's body is ill suited for the perception of smoothness and roughness, hardness and softness, or for any sort of analytical touching at all; because of the lever-like action effect of the hairs and feathers, it is probably suited for adroit movements in space and for the recognition of stimulation sites on the animal's own body. For example, I am always struck by how well horses can localize the point at which a fly is sitting or where one has very gently touched their hair ends. Hairs and feathers, and even more so the keratinized portions of the extremities, act to screen out nearly all thermal qualities of the objects touched. The extremities of mammals, except for those that have grasping hands, also are not suited for stereognostic tasks; for this, the mouth is better suited.[7]

Animals with bare skin, such as frogs, earthworms, and planaria, have been shown to discriminate roughnesses. "Many crayfish seek a rough support, so they must be able to recognize it as such (Kafka, op. cit., p. 37)."[8] Tactual sensations produced by mutual contact provide bees the means for communication during their courtship dances, according to the beautiful work by von Frisch.[9] The mutual trilling of ants can also be mentioned here.

We conclude these comparative psychological considerations with a step into the world of plants. "The botanists distinguish a sensitivity to sharp blows from a sensitivity to contact stimuli. Plants that are sensitive to jolts (the best known example is the mimosa) react with full force of movement even to a single impulse, whereupon a relatively long refractory period . . . follows."[10] Thus, our theory, which sharply distinguishes the two branches of the sense of touch, the pressure (contact) sense and the vibration sense, holds even in botany.

Footnotes

1. K. von Frisch, Ein Zwergwels, der kommt, wenn man ihm pfeift (A dwarf whale that comes when one whistles), *Biol. Zentralbl.*, *43*, 1923.
2. G. Kafka, *Tierpsychologie* (Animal psychology), Munich, 1922, p. 38.
3. Ernst Matthes, Die Rolle des Gesichts-, Geruchs- und Erschuetterungssinnes fuer den Nahrungserwerb von Triton (The role of the senses of vision, smell, and vibration for acquisition of food in tritons), *Biol. Zentralbl.*, *44*, 1924, p. 83 f.; and by the same author: Das Geruchsvermoegen vom Triton beim Aufenthalt unter Wasser (The underwater olfactory ability of the triton), *Zeitschr. f. vergl. Physiol.*, *7*, 1924.
4. Haecker, Ueber Lernversuche an Axolotln (Learning experiments with axolotls), *Arch. f. d. ges. Psychol.*, *20*, 1899.
5. According to F. Baltzer, many spiders likewise localize the center of vibration caused by a captured buzzing fly in the web sufficiently precisely to find the prey based on this. *Die Naturwissenschaften*, *12*, No. 45, 1924.
6. I suspect that there is a connection here with the recent, remarkable findings of O. Klemm, Ueber die Wirksamkeit kleinster Zeitunterschiede auf dem Gebiete des Tastsinns (Effectiveness of very small temporal differences in the domain of the sense of touch), *Archiv f. d. ges. Psychol.*, *50*, 1925, that the impression of directionality occurs during tactual stimulation of symmetric parts of the body (both index fingers) if there is an interval of a few milliseconds between the two stimuli.
7. On this, see L. Edinger's statements on the oral sense in his lecture on animal psychology. *Bericht ueber den 3. Kongress f. exp. Psychologie* (Report on the Third Congress for Experimental Psychology), F. Schumann, Leipzig, 1909.
8. New material that is relevant here is also contained in the recent paper by H. Balss on adaptation and symbiosis in pagurides. *Zeitschr. f. Morphologie u. Oekologie der Tiere*, *1*, 1924.
9. K. von Frisch, *Ueber die "Sprache" der Bienen* ("Language" of the bees), Jena, 1923.

10. M. von Frey and F. Kiesow, Ueber die Funktionen der Tastkoerperchen (Functions of the tactile bodies), *Zeitschr. f. Psychol.*, *20*, 1899, p. 159.

Section 44. The Sensory Organs of the Vibration Sense

The justification for speaking of a vibration sense is based on the special status, phenomenologically and functionally, of the vibration sensations compared with the pressure sensations, the only sensations with which they could be grouped. In general, to speak of a separate sense requires evidence of specialized sensory organs. Such evidence is not presently available for the vibration sense. However, it should be noted that the assignment of the other cutaneous senses (pressure, warm, cold, and pain) to the sensory organs found in the skin has scarcely passed beyond the stage of speculation either. The great profusion of nervous elements in the skin almost requires that even more senses be distinguished.[1] We find structures in the skin that, by their size and other characteristics, seem hardly less suitable for transmitting oscillations than the elements of the basilar membrane, which in Helmholtz's theory are taken to be resonators for tones.[2] This is not to say, however, that Ewald's pressure-pattern theory cannot perhaps do even better at helping us to understand the functioning of the sensory organs for vibration.[3] The deaf learn through methodical instruction to perceive and discriminate very diverse frequencies of oscillation with the fingers. Are we really to suppose that the fingers, and the other portions of the body of comparable vibratory capability, are outfitted with organs as complicated in structure as the Helmholtz resonance theory would require? Does Ewald's theory not offer a better possibility of obtaining an explanation? That theory, it seems to me, makes it easier to understand why the upper and lower frequency thresholds of oscillations differ so much for different positions (as may be concluded from the data of all the authors who have concerned themselves with vibration sensations). The greater the range between the two thresholds, the more capable we must consider the vibratory sensory organ to be. In a sense, our body provides a view of the evolution of these organs, from the simplest (with the narrowest range) to the most advanced (the ear, with the widest range). I rather think that the considerations just mentioned can breathe new life into Ewald's theory of hearing.

If Wundt[4] can say that "a pattern emerges in numerous insects, which seems to unify the three evolutionary stages of organs of touch, tension, and hearing," and thereby actually coordinate three senses in the same sensory organ, then why cannot the pressure sense and the vibration sense be coordinated in the same sensory organ in humans? It is quite conceivable theoretically that devices found in the same sensory organ make it sensitive to two different forms of stimulation and produce, accordingly, two different types of sensations.[5] Von Frey recently defended such a view while, as already mentioned above (Section 39, Subsection 1), developing his own position in various publications.[6] "Research in recent years has shown that the cutaneous pressure sense transmits four types of sensations, which are designated as contact, tickling, buzzing, and pressure." Here we are only interested in the two sensations, pressure and buzzing (vibration), which, although admittedly qualitatively distinct, von Frey assigns to the same sensory organ. Why, von Frey asks, are these sensations not readily recognized as being very closely related, indeed, as essentially the same, and as only different in form? Why do we not get the immediate impression that the same basic process underlies them both, and that the differences only reflect the different spatial and temporal patterns of excitation of the receptors? "The answer can only be that the excitation in the peripheral receptors is not unaffected by where it is that the physiological (psychophysical) processes corresponding to the conscious impression occur. The excitations interact with each other, fuse to a certain extent, and form new complexes." Without finally deciding the issue, it seems to me more probable, as I have said elsewhere (Katz, 1923a, 1930), that there are no special sensory organs for the sense of vibration, and I would therefore agree with von Frey in this case.[7] This judgment, it ought to be made clear, as von Frey was also aware, allows one to reach a conclusion that contradicts a basic theoretical tenet of sensory physiology. "As is readily apparent, a sort of partial breakdown of the theory of specific nerve energy is inherent in the possibility of evoking qualitatively different sensations through stimulation of one type of receptor."[8] I draw this conclusion for myself, because I regard the vibration sensation as involving a separate sensory modality in Helmholtz's sense, and not a modified pressure or contact sensation. Strictly speaking, von Frey did not need to draw this conclusion for himself, because the qualitative difference he found between pressure and vibration sensations constitutes no more of a breakdown in Johannes Mueller's theory than does perhaps the finding of qualitative differences in the domains of tones or colors.

M. von Frey sees vibration sensations as representing forms or gestalten of the pressure sense in line with Gestalt theory. He maintains that they correspond to Koehler's definition, by which gestalten are those states and processes whose features and effects do not represent a simple sum of the features and effects of their so-called parts. I do not deny the fruitfulness of Koehler's notion where it is applicable, but fail to see how it could be applied to the vibration sensation. No parts at all can be distinguished in the case of the vibration sensation; it exists only as a whole. Its intensity can be varied, and its duration can be reduced down to that of the vibration impulses, but these variations are completely different in nature than those found when a real gestalt, such as a visual figure or a melody, is analyzed into its elements. It is as proper (or improper) to speak of gestalten in the case of vibration sensations as in the case of tones. I trust that with individual tones no one would find it very easy to speak of different gestalten formed from the elementary air-pressure fluctuations. All genuine gestalten are considerably influenced by the accompanying interpretation; this is shown by every visual figure, by every melody. This fact leads to another, namely that individual experience makes its particular contribution to each gestalt. We learn by practical experience to name a particular tone or to bring it into one relationship or another with a second tone, but the tone phenomenon itself does not thereby undergo any real change. Except for the external associations it evokes, a particular tone is influenced relatively little by individual experience. I would now like to assert that what has just been said about tones holds equally well for vibration sensations: they are influenced by practical experience in the same way and to the same extent—and only to that extent—as the tone sensations.

If a phylogenetic and far-reaching functional relationship is confirmed between audition and the vibration sense, should a phenomenological similarity between their sensations be claimed as well? I would only answer for now that if forced to judge, I would choose the series (pressure sensations - vibration sensations - auditory sensations) as expressing the existing phenomenological affinities.

Footnotes

1. "Few organs in the human body are so profusely and diversely equipped with nervous elements as the skin," A. Jesionek, *Biologie der gesunden und kranken Haut* (Biology of healthy and unhealthy skin), Leipzig, 1916.
2. According to M. von Frey and F. Kiesow (op. cit., p. 149), there are very significant osmotic and frictional resistances in the human skin, with its 75% water content, so that, for brief and weak deformations, fluid movement can be disregarded and the skin can be regarded as completely elastic. Naturally, this makes it particularly suited for picking up vibrations.
3. On this, see R. Ewald, Eine neue Hoertheorie (A new theory of hearing), *Pfluegers Archiv*, 76, 1899. Ewald conveyed the quintessence of his theory, whose basic thrust better harmonizes with the wholistically-oriented present-day psychology than with a resonance theory built upon the elements, in the following sentence: "In the ear, the impulses from the sound produce on the basilar membrane a wave pattern (sound pattern), whose special form enables the basilar membrane to become a link in the transmission chain between the sound and sound sensation."
4. W. Wundt, *Grundzuege der physiologischen Psychologie* (Elements of physiological psychology), 6th ed., 1908, Vol. 1, p. 442. I should also like to mention an observation here that deserves special consideration in light of the citation above from Wundt. If you place a vibrating tuning fork on your forehead, then an un-

pleasant experience bordering on vertigo usually occurs. The organs of balance and equilibrium are obviously excited by the vibration.

M. Ettlinger, in the *Tierpsychologischen Anmerkungen* (Remarks on animal psychology) cited above, pointed out the clear progression of stages, from the seismographic organ to the organ of hearing (demonstrated by Hensen's experiment with a crayfish, whose stripped-off auditory hairs respond specifically to different tones).

5. "The fact that pressure and vibration sensations, even on the same part of the body, can be perceived simultaneously causes no problem, since two different types of stimuli acting simultaneously on the same organ also can produce coexisting phenomena; for example, the same vibration can result in a smooth, binaural tone and a monaural beat." E. von Hornbostel, lecture.

6. M. von Frey, Die vier Empfindungsarten des Drucksinnes (Four types of pressure sensations), *Zeitschr. f. Biologie*, 79, 1923. Ueber Wandlungen der Empfindungen bei formal verschiedener Reizung einer Art von Sinnesnerven (Transformations of sensations with formally different stimulation of one type of sensory nerve), *Psychologische Forschung*, 3, 1923. Kitzel-, Beruehrungs- und Druckempfindung (Tickling, contact, and pressure sensations), *Skandinavisches Archiv f. Physiologie*, 43, 1923. Felix and M. von Frey, Versuche ueber den Hautkitzel (Experiments on tickling of the skin), *Zeitschr. f. Biologie*, 78, 1923.

7. Further support for this position comes from the fact that the cornea and conjunctiva of the human eye, which lack contact and pressure sensations, also fail to respond to tuning-fork stimulation with vibration sensations. M. von Frey and W. Webels, Ueber die der Hornhaut und Bindehaut des Auges eigentuemlichen Empfindungsqualitaeten (Sensation qualities characteristic of the cornea and conjunctiva of the eye), *Zeitschr. f. Biol.*, 74, 1922.

8. In an investigation, Ueber die Sensibilitaet, insbesondere den Drucksinn, vom physiologischen Gesichtspunkt aus (On sensitivity, particularly that of the pressure sense, from a physiological point of view), *Klin. Wochenschr.*, 2, No. 46, 1923, V. Baron von Weizsaecker emphasizes that a certain breadth must be given to the concept of specific nerve energy if it is not to lead us into error.

Section 45. Perception of Object Properties Through the Vibration Sense

Section 43 dealt with the perception of vibratory *states*; we now turn to the perception via the vibration sense of the *properties* of the objects in our environment. I am no more accustomed, as a layperson, to interpret the noises produced by driving a railroad engine or operating a machine as reflecting the permanent properties of these objects, than I am to do the same for the vibrations that may reach me in these cases. The auditory or vibratory information is not distinctive enough for me to handle such a task. It is a different story in the following case: I cannot decide what the material is in an object presented to me, so I tap it with the edge of my fingernail, and then the desired information is provided by the resulting noise to the ear and by the resulting vibrations to the sense of touch. For the time being, we will completely disregard the auditory aspect of the process, and by stopping up our ears, we again limit the experiment as far as possible to the tactual aspect. The material reveals itself to the touch organ. I discover whether I am dealing with an elastic or inelastic material, a relatively hard or soft material; materials can even be discriminated that give exactly the same impression in "superficial" touching. The information is provided not by the typical touching process lasting a few seconds, but by a *vibration impulse* lasting a slight fraction of a second.

I carried out several experiments on the duration of a vibration impulse that can usually lead to recognition of a material. An iron thimble is fastened on one finger and at the same time connected by an easily movable wire with an electric battery. The elastic nature of a metal plate is recognized from the vibration impact when it is struck very briefly with the thimble. At the instant of contact between the thimble and metal plate, an electrical circuit is closed. The duration of the contact can be recorded graphically or with the help of a chronoscope. The briefest contacts were about .01 sec. Since the excitation of the touch organ does not begin immediately upon contact between thimble and plate, nor cease upon their separation, the time we have measured does not state exactly the duration of the vibration impulse, but ought to agree well with its relative order of magnitude. It also seems reasonable to assume that putting on the thimble in order to achieve a good closure of the electrical circuit does not really change the duration of the vibration impulse that would be obtained by striking the plate with the bare fingernail. Likewise, in working with a hammer or some other tool, the character of the vibration impulse permits us to recognize the material that is being struck by the tool. Accordingly, it was of interest to study the durations for such cases as well.[1] The duration of hammer-iron plate contact decreased to .005 to .003 sec (!),[2] and yet there was confident, reliable recognition of the material.

This can only mean that the object struck by a fingernail or tool is thereby placed into a vibratory state characteristic of it, which is then impressed on the vibration sensory organ. Recognition of a specific material, such as wood, cardboard, metal, porcelain, etc., is less common by far than simply being able to discriminate the materials. This is not surprising, considering that normal persons do not pay the attention required to fully exploit this tactual information channel. Hearing persons get far more practice in using the noise produced by the blow to recognize the material; for this reason, the extraordinary diversity of the noises produced in the course of such experiments can serve as a means of control for us.

Vibration impulses also provide information on the nature of the floor that we set our feet on or touch with the end of our cane. Except for a few transparent materials, the eye pales before the vibration sense, which enables us to fathom the inner nature of objects. The properties of things reveal themselves in a distinctive manner to the vibration sense; the information is dynamic-active as defined above (Section 42).[3]

Vibration impulses can be perceived by virtually every part of the body. Even if one taps a small hammer only weakly on a table top, a second person can still feel these blows at a relatively great distance with the hand, elbow, or clamped teeth. Strange to say, the touching does not work with the tongue, a sure sign that vibration sensitivity does not proceed simply in parallel with pressure sensitivity and the precision of the spatial sense. Vibration impulses very likely are also components of the pulse; this may be related to Hausmann's observation, cited above, that the tip of the tongue cannot feel the radialis pulse.[4]

The vibration sense participates in the act of touching in yet other ways and to a far greater extent than depicted above, and we now present a systematic account of matters already touched upon above several times. Put an ear to the touched surface while executing touching movements with the fingers, and you can confirm that continuous noises arise, which differ more or less from one another in intensity and quality. After some practice, the different papers in a set can be discriminated quite well based solely on the noises they produce.[5] These perceptible noises are an unmistakable sign that with movement of the finger, vibrations arise at the point of contact with the tactual surface, and it now will be contended that only the vibration sensations thereby evoked permit finer discrimination of the tactual surfaces based on their roughness or smoothness. The smoother the touched surface, the weaker the vibrations, but whether the vibrations disappear completely with the smoothest surfaces remains an open question. I have yet to encounter a surface that allowed not even a little noise to be heard near it during normal touching, and on which no vibrations occurred. This alone is naturally not crucial, because it is a separate question as to whether very weak vibrations are still able to excite the sensory organs. With very smooth surfaces, introspection reveals nothing about vibration sensations, but we should not place too much stock on an introspection that has been so little tutored in these sensations. It is conceivable that the vibration sensations are completely absent with very smooth surfaces, just as they likewise vanished in the experiments with glue described above (Section 24, Subsection 5), and that precisely this absence, along with the retention of contact sensations, characterizes the smoothest surfaces. With rougher surfaces, it is easy even for the psychologically unpracticed to confirm the occurrence of vibration sensations. Move the edge of a fingernail over a very rough surface, then the vibrations are felt not only in the fingernail, but in the entire hand, and even as far as the forearm. This shows how considerable is the resonance of certain portions of the human body to the vibrations produced in this case, as has already been pointed out.

What are the implications of the hypothesis[6] presented here that the vibration sensations are the essential component in the judgment of roughness? It has already been amply demonstrated above that the smallest differences in level on our test papers were too small to be recognized as such by the pressure sense; we contend that they are apprehended by means of the vibration sensations that arise from rubbing. The efficacy of the vibration sensations is even more clear when the papers are discriminated, with no impairment in performance to speak of, through collodion and adhesive tape, that is, with a tremendously increased pressure threshold, or even at a distance using a rod. Recall again the result described above that most differences between papers are recognized even when the papers are no longer touched in the usual way with the volar side of the fingertip, but with the dorsal side, that is, with the fingernail. Naturally, we cannot speak of a literal, iconic transmission of the gestalt of the paper surface to the skin lying beneath the collodion or adhesive tape, or to the fingers holding the rod, nor of a corresponding deformation of the stiff fingernail and its transmission to the pressure-sense organs underneath. The result of these experiments can be readily understood in terms of our hypothesis. Vibrations quite certainly arise in rubbing the paper with the collodion or adhesive tape, the rod, and the edge of the fingernail, and are transmitted through these intermediaries to excite the sensory organs in the skin.

The role of the vibration sensation in producing distinctive surface impressions explains many an oddity of the experiments presented in Sections 18 to 26. It was shown that the discriminability of the papers tested suffers little from the introduction of intermediaries (collodion, adhesive tape, wooden rod), but that all papers feel completely different when touched with intermediaries rather than with the bare finger, and therefore comparisons of papers with and without intermediaries are very difficult to make and produce uniformly poor results.[7] Our hypothesis would well-nigh predict that type of result. For physical reasons, completely different vibrations arise when, instead of the bare skin, collodion or adhesive tape is used, or when the wooden rod moves over the papers (as the monitoring ear will readily attest), producing completely different vibration sensations and thereby different surface impressions as well. The discriminability of the papers need not thereby suffer, so long as the vibrations obtained via intermediaries, even if not the same, remain just as distinctive as those obtained using the bare fingers. The comparison of papers that are touched with and without intermediaries, however, must become very difficult. It is not clear to what extent the small reduction in discriminability observed with intermediaries (collodion, adhesive tape, wooden rod) is due to the fact that the vibrations at the point of friction differ somewhat less, or to the fact that the intermediary has a damping effect on the transmission of vibrations to the sensory organs and thereby an equalizing effect. I consider the second factor to be more significant. If one uses a medium that muffles the vibrations completely or nearly so (felt, cloth), then, as already mentioned, the discriminability of the tactual surfaces is completely eliminated. If, as in the experiments in Section 29, the tactual surfaces are placed on a balance plate that moves immediately as soon as the pressure of the touching finger exceeds a certain low value, then the discriminability of most surfaces is considerably reduced, although the experience of resistance remains in every case. Evidently, the physical conditions no longer permitted sufficiently great friction to produce vibrations.

If one tests the discriminability of our papers or other material at other positions on the body besides the fingers, then the performance is found to fluctuate much less than would be expected based on differences in the pressure threshold and the stereognostic ability of these positions. If, as shown in Section 25, the large toe does scarcely less well than the finger in discriminating the papers, it is because its sensitivity to vibrations is just about as high that of the finger. As we already said above, this finding indicates that the basic structure of a sensation is largely independent of practical experience, which agrees with the view developed here concerning the vibration sensation. The fact that touching with the teeth is based on vibration sensations was already anticipated above. Recall also that amputees discriminate tactual materials with their upper arm stumps not that much more poorly than with the fingers, although the pressure threshold is much higher at the stump and the stereognostic ability is minimal. The discrimination is based primarily on the vibration sensations evoked at the stump, whose sensitivity to vibrations is not so very far behind that of the fingers. Likewise, the discrimination of tactual materials by means of prostheses, insofar as this is yet possible at all, can only be understood in terms of vibrations transmitted from the friction point of the prosthesis to the stump.

If the vibration sensations play the role ascribed to them here, then it becomes clear why the movement of the touch organ with respect to the tactual surface was of such great significance in all of the experiments mentioned. Only frictional movements produce the vibrations that evoke the vibration sensations necessary for discrimination. The surmise that surface roughness varies with the speed of the touching movement was confirmed by the research in Section 23. We found there that the perceived character of the tactual surface was substantially altered only at extremely slow and extremely high speeds, evidently due to the concomitant change in vibrations with speed of touching. The variation in the vibration sensations was obvious across a rather broad range of intermediate speeds, but the judgment of surface smoothness or roughness varied scarcely at all. It is apparent from this, that within certain limits, the speed of movement (and concomitant change in the vibration sensations) is taken into account to a remarkably precise degree.

In my experience, wire-wound rods are particularly suitable for the exact study of this effect; with these rods, one can control the decisive variables. I wound a round pencil with a 10-cm length of .25-mm thick, spun copper wire. The resulting variation in height between successive ridges was indeed so considerable as to be apparent in its general form or gestalt even to the spatial sense of the skin with suitable touching. If one moves a finger over the pencil with some speed, however, then the impression of a certain "regular" roughness occurs for the vibration sense. If the speed is set very slow, about 1 cm/sec, so that about 40 windings per second pass by one point of the finger, then the successive elevations feel like the elements of a string of pearls moved on the finger. At speeds above 1 cm/sec, the experience is of regular roughness. The speed can be increased about 10 fold without really changing the judgment of roughness, but at yet greater speeds, to be sure, the impression becomes smoother and finally intensifies to that of painful contact. What should be of prime concern to us is the fact that the roughness judgment remains the same at intermediate speeds. In increasing the speed from about 1 to 10 cm/sec, the number of individual windings exciting the skin increased from 40 to 400. This increase is almost completely nullified by taking into account the speed. To preclude errors, the compensation process must involve not a conscious inference, but operate in a similar way and, at the limits, with similar constraints, as the processes that take into account illumination in the case of color [and brightness] constancy, and distance in the case of size constancy. Variation in tactual speed and the concomitant vibration sensations can be taken into account not only with surfaces of regular roughness, like the wire-wound rods, but also surfaces with irregular roughness. To this must be added the fact that the variation in the vibration sensations resulting from variation in the pressure intensity of the touch organ likewise is compensated for, within certain limits, by taking into account this pressure intensity.

The use of wire-wound rods also is more suitable than other types of procedure for the investigation of certain parallels between auditory and vibratory impressions. Move your finger over the wire winding at different speeds, then the ear bears witness to the fact that completely different vibration states are induced. The greater the speed, the higher the pitch,[8] and if the speed is varied continuously, a vibratory glissando occurs, which rises with increasing speed, and falls with decreasing speed.[9] The occurrence of this glissando is reminiscent of the Doppler effect. Whereas a vibratory glissando does undergo a compensation, because the variation in movement that produces it enters into consciousness and is essentially taken into account, an acoustical glissando that occurs according to the Doppler effect is well known not to undergo a compensation; thus the sound of the moving source is not apprehended as a tone of constant pitch. The functional relationship between sound frequency and the speed of movement of the sound source is not comprehended in precisely the way required for compensation, presumably because of insufficient experience.[10] We have advanced to a point where the parallel between the auditory and vibratory senses no longer holds. The vibration sense proves to be the more resilient sense, the one better adapted to the stimulus material.

To those who question the compensatory principle just presented for vibration sensations because it is too complicated, let me point out that other models could not handle the facts under discussion without such a principle. Ascribing the impression of roughness to pressure sensations or to resistance sensations in gliding, rather than to vibration sensations, does not escape the need to allow for a far-reaching taking into account of how these types of sensations vary, depending on the speed of touching movements, which is no less mysterious here by any means.

I do not attribute all roughness judgments to vibration sensations, but call upon the latter only for the levels of roughness that are no longer apparent to the spatial sense of the skin, as well as for the cases where the presence of intermediaries precludes deformation of the skin. Once the differences in surface level exceed a certain threshold, they are also recognizable to the sensory organs for pressure (first if moving, and finally even if motionless). The level of this threshold cannot be specified with complete certainty, but it clearly has already been exceeded with coarse fabrics. When the touching hand recognizes the type of weave, in addition to the general character of roughness or smoothness of a woven fabric, then the activities of the vibratory and pressure senses have been merged. In my experience, vibration sensations contribute less to the impressions evoked by touching material composed of easily movable particles (sand, flour, sugar, etc.). The physical conditions disfavor the generation of vibrations; the particles have a greater effect on the pressure sense.

The work with wire-wound rods showed the impression of roughness to be little affected by whether the rod is moved over a very small or a very large portion of the finger, or by whether the rod is moved over only one finger or over several fingers at the same time. This is not difficult to understand theoretically, since only the quality of the vibration sensation matters, and that in turn depends on the magnitude of friction between the touch organ and the tactual surface. One could well imagine that the impression of roughness depends fundamentally on the vibration state of a sensory element; it would be more difficult to conceive how the spatial sense of a tactual point could provide us with an appropriate representation of the roughness of a tactual surface. This explains not only the unitary impression of surface roughness obtained when touching with five fingers, but also why a very large reduction in the size of the tactual surface impairs its discriminability only very slightly (Section 20).

We introduced above the distinction between static-passive and kinematic-active sensory information. The rough and smooth tactual properties of the object reveal themselves to us in a dynamic-active way.[11] Roughness is not a quality of the vibration sensation. We can only say that a vibration sensation can help to establish the roughness property of an object. To a certain extent, the vibration sensation provides a raw material, which leads either to the judgment of a state (Section 43) or to the judgment of an object property (this paragraph). The "how" of the latter process will be briefly considered. Vibrations *never* mediate the impression of a tactual surface as such, no matter how small the surface. In order for a vibration sensation to be referable to a tactual surface, the latter must have been established in another way. The predominant view is that the spatial sense of the skin creates the surface image, with the principal participation of the pressure sense. This view can hardly be maintained as such; it needs to be amended in line with the statements in Section 46 showing the visual involvement in the spatial tactual structures. Regardless of whether one advocates the original view or the amended version, however, the tactual surface is based on the static-passive senses, and only this tactual surface provides the dynamic vibration sensations with the properties we have dealt with in detail. The structure of this process is quite remarkable.

As reported in Section 42, inserting papers of different roughness between a motor-driven wooden disk and a wooden rod produced corresponding changes in the vibration sensations obtained from the rod. However, since a "touching on the spot" occurred here, the vibrations were apprehended only as appertinents, not as properties of objects. It appeared that by returning to the experiment we might artificially induce to some extent the impression of a rough surface, where in actuality the surface is smooth. Within certain limits, this plan was successfully executed. A firmly-clamped glass plate was set into vibration. When I move my bare finger over it, then I feel a vibrating surface of smooth glass. If I touch the glass with a little rod, however, then the illusion of a rough surface comes through quite clearly. The intervening medium is needed to eliminate the tactual impression of a finger resting on the glass, which would destroy the illusion. Experiments on these questions are still in progress.

Footnotes

1. We also notice whether we strike the blow with a wooden or an iron hammer, even when their having the same weight and distribution of weight leads to the same expenditure of effort in both cases. The rebound is a painfully intensified vibration impulse.
2. At such short intervals, the sound quality naturally cannot be clear, but only noisy.

3. A very remarkable observation from comparative psychology will be interjected here. A. Hase (*Die Naturwissenschaften, 12*, 1924, p. 380) describes as follows the behavior of the female ichneumon-fly, *Lariophagus distiguendus*, when it pierces through the cocoon of the larvae of another ichneumon-fly (*Habrobracon juglandis Ashmead*) in order to lay its eggs: "The . . . feelers tap at a very fast tempo on a small spot located directly beneath the female's body. This vibrating touching and knocking can best be compared to playing a trill very fast on a piano." The carefully selected spot is then pierced through. The behavior permits scarcely any other interpretation than that the vibration impulses produced by the trilling are supposed to provide information on the thickness and other properties of the spot to be breached on the cocoon. Thus, palpatory examination on the basis of vibrations is performed by an insect!
4. Goldscheider and Hoefer (op. cit., p. 316) carried out experiments in this regard with pulsating tubes.
5. One establishes in this instance how mighty the realm of noise is.
6. Occasional spontaneous utterances that interpret vibratory and auditory impressions tactually also support our hypothesis. Thus, a deaf person says that when a cymbal is beaten, "I feel as if I were touching silk (Eschke, op. cit., p. 43)." In a study by E. M. Edmonds and M. E. Smith, one observer described the octave as "smooth, like the touch of polished glass," and the major second as "gritty, like sandpaper." The phenomenological description of musical intervals, *American Journal of Psychology, 34*, 1923, pp. 287-291. Buerklen (op. cit., p. 79) provides a summary of those auditory and tactual impressions that appear related to the blind.
7. It is worth mentioning in this connection that a finger numbed by the cold or in some other fashion is called deaf (*taub*) in German and mute (*njem*) in Russian.
8. If the finger moves over the rod at a constant speed, then the impression of a *noise* predominates. If, however, it moves over the rod at rapid but varying speeds, then more of a *tone-like quality* emerges for each individual noise, due to the spontaneous mutual comparisons that then occur. The greater the speed, the higher the pitch. Using the fingernail makes the tones especially clear. G. Révész, *Zur Grundlegung der Tonpsychologie* (Foundation of the psychology of sound), Leipzig, 1923, p. 72 f., thoroughly covered sound phenomena of the sort associated with noise, and their explanation.
9. As a rule, the rising (sinking) glissando occurs as a crescendo (decrescendo), since the finger pressure spontaneously increases (decreases) at a greater (lesser) speed.
10. This point ought perhaps to be restricted to the sighted. According to Truschel's careful observations, in the course of auditory localization, many blind persons appear to take into account the change in pitch as a result of movement. "In walking to one side past a relatively large object, the noise of the steps increases in pitch at that moment when one is opposite to the object, and then remains constant until one has left the object behind. On the other hand, when one approaches objects crossing his or her path, the pitch increases steadily." W. Steinberg, *Die Raumwahrnehmung der Blinden* (Spatial perception in the blind), Munich, 1920, p. 38.

11. We say "the material is rough" (static), but we can also say [in German] that "the material roughens" (kinematic). With colors, likewise, in addition to the usual static form of expression, such as "the meadow is green, the heaven is blue," there is also the poetic kinematic, "the meadow greens, the heaven blues." Characteristically for audition, however, we can use only the kinematic form of expression, "to sound."

Addendum. The Auditory Recognition of Object Properties

As a rule, noises provide clues as to *processes* in our environment, but now and then they also permit inferences about *properties* of objects. The sighted can usually pick up the object properties in a quick glance, after having become aware of them auditorily. It is different with the blind, who glean much more from auditory impressions. "If different plates are thrown onto a table, then a blind person can not only tell what shape they have, but also what metal or wood they are made of."[1] The blind person cited above (Cohn, op. cit., p. 76) stated that auditory impressions are very informative in recognizing object properties; with their aid, a blind person recognizes parcel-post wagons, furniture wagons, rack wagons, and taxis. The sighted, if they so desire, can pay greater attention to these things, and thereby improve greatly in this respect. I am reminded how experts infer the properties of a material they are constantly dealing with from its sound, just as machine operators recognize from deviations in the running noise that something has changed in the machine and where it has changed. It is noises, not tones in the musical sense, that prove to be of such practical value in these and other cases. The tone seems like a luxury, which has less biological significance than noise. Vibration exhibits no luxurious tendencies (if we disregard for now such singular cases as that of Sutermeister). It shows us a sober stage in the development of a sense in which everything is still based on practical usefulness. The auditory sense also has probably had to pass through this stage. Thus, the comparison with vibration helps us to see audition from a perspective that has not always been sufficiently appreciated.[2]

Footnotes

1. F. Hitschmann, Ueber Begruendung einer Blindenpsychologie durch einen Blinden (Foundation of a psychology of the blind by means of a blind subject), *Zeitschr. f. Psychol.*, 3, 1902, p. 391. F. Zech likewise called attention to the use by the blind of auditory impressions produced by scratching with the fingernail. Blindenschule, 1919.

2. L. Bard appears to present a similar view of audition. "The foundation of auditory perception is not the perception of tonal quality, but rather, just as in vision, the recognition of the shape and properties of external objects. The sensory patterns have physical elements, which provide the objective basis for perception. While vision permits the recognition of the external shape, audition mediates the deeper properties (the linings, hollow spaces, inner contents of the objects)." *Arch. internat. de laryngol., oto-rhinol., 1*, 1922. The citation above is taken from a report in the *Zentralbl. f. d. ges. Neurol. u. Psychiatrie, 32*, p. 340.

Editor's Notes on Chapter II, Division III: The Contribution of Visual and Kinesthetic Processes to the Structure of Tactual Forms (Sections 46 and 47)

In the preceding chapter on the vibratory sense, Katz discussed the close ties touch has with audition. In the next chapter, Katz discusses the ties touch has with vision and the kinesthetic sense. Touch, vision, and audition are exteroceptive senses, whereas kinesthesis and the vestibular sense (sense of balance) are proprioceptive senses (Geldard, 1953). The kinesthetic sense has receptors in muscles, tendons, and joints, but the joint receptors are now regarded as the most important. Katz emphasized the muscle component and the experience of effort in Section 47, but the position information provided by the joint receptors may also have made an important contribution, even in the perception of movement. The independence of the tactual impressions from the force of movement indicates, according to Katz, that the kinesthetic impression was taken very precisely into account. In Section 17, Katz cited a similar constancy, in which the change in position of the hand while touching a motionless object was taken into account. In that section, Katz also mentioned another source of position information, that provided by the centrally-triggered motor commands which direct the hand movement (see also Weiskrantz, Elliott, & Darlington, 1971).

In Section 46, Katz presents a mixed picture on the role of visual imagery in touch. Visual imagery permeates touch, and its effect seemingly persists even when the eyes are closed, so congenitally blind subjects are needed to determine what qualities touch possesses initially. What little data Katz had access to (mainly Goldstein and Gelb's), indicated that vision was absolutely crucial in one respect, that of providing the quality of spatiality to the tactual representation. This point was later emphasized by Katz's colleague, Révész (1950). Vision surpasses haptics, according to Re've'sz, mainly because it examines objects in a simultaneous fashion, rather than in a successive or piecemeal fashion. "The more complicated the tactile object, the more difficult is the haptic apprehension of proportions and the more marked becomes the superiority of the visual . . . sense" (p. 141). Seeing is indispensable to the sculptor, he noted, "for only vision is capable of raising the sensory impression into the sphere of aesthetic contemplation" (p. 328). All of the blind sculptors Révész discovered were late-blind; "this fact and its converse—the fact that those blind from birth have never produced any plastic work of aesthetic value—leaves no doubt whatever as to the importance of visual impressions as the basis of tactile experience" (p. 234). Other research likewise indicates that vision surpasses haptics in two-dimensional spatial perception, both in encoding small-scale manipulatory displays and for cognitive mapping of large-scale, ambulatory space (see review by Lederman, Klatzky, Collins, & Wardell, 1987). Whereas a single fixation or a quick scan suffices for visual perception, "the manipulatory and ambulatory systems commonly gather information by a sequence of relatively slow exploratory contact movements over surfaces and along contours" (Lederman et al., 1987, p. 607), thus imposing heavy demands on memory and temporal-integration processes.

Chapter II. The Contribution of Visual and Kinesthetic Processes to the Structure of Tactual Forms

Section 46. The Influence of Visual Representations

1. *Observations on sighted persons.* In Section 4, it was primarily with a heuristic motive that we contrasted visual and tactual perception as to their general structure and at many individual points, emphasizing this or that component property. We hoped that by starting from vision as the better-known field, we could more readily trace the corresponding, or even markedly deviating, new facts in the sense of touch. Having already alluded to the theme above (Section 29, Subsection 6), let us return once more to it here. We now examine the interrelationship between touch and vision, and the interweavings into which they enter. In particular, we seek to determine what contribution visual processes make to the fashioning of tactual forms. Which tactual images contain a visual component, and which ones do not and thus ought to be characterized as tactually autochthonous, i.e., rooted in the tactual domain according to sense and gestalt?

Since John Locke's discussion in his *Essay Concerning Human Understanding* of the Molineux problem (can a congenitally blind person, having learned to discriminate cubes from spheres by touch, discriminate at once the two objects by sight as well, when made to see?), many people have occupied themselves with the questions raised by this problem concerning the relationship between vision and touch. Successful operations on the congenitally blind by ophthalmologists have provided an answer to Molineux's problem that, despite its ambiguity on more than one point, basically confirms Locke's proposed solution, which involved external associative links between qualitatively quite different visual and tactual impressions. Not only is the information obtained from this side in the empiricism vs. nativism controversy not unequivocal, but this dispute itself is completely irrelevant here, since only the interrelationship between the two senses is to be studied. In his thorough investigations into the relationship between the separate tactual and visual experiences, W. Steinberg tried to make clear "why, even for the congenitally blind, phenomenal three-dimensional objects within the narrower and broader confines of tactual space resemble those provided by the eye in the sense that in both cases the structure of the stimulus is adequately expressed."[1] If he is correct, then it is virtually a foregone conclusion that visual interpretations are made of tactual impressions, just as the tactual tinge lends meaning and significance to many visual impressions. Goldstein and Gelb (op. cit., p. 3) cite Parish, Pillsbury, Washburn, Henri, Judd, and Churchill as bearing witness to the effect of the goodness of the visual image on tactual performance. We wish to add to the list the names of Bonaventura,[2] Fitt,[3] von Frey,[4] Jaensch,[5] Rupp,[6] Spearman,[7] Ziehen,[8] and Gellhorn,[9] without thereby making any claim of exhaustiveness for the list. The work of Ahlmann[10] is particularly fine, because it comes from a blind person who was very well trained psychologically. The experience of everyday life seems to show that visual images are aroused more readily by tactual impressions than vice versa; experimental data confirm this.[11] The concurrent evocation of tactual imagery during the presentation of pictures has been demonstrated by Martin.[12] Weber was concerned with the estimation of tactual magnitudes. "In judging distance, we do not rely (as sighted persons) on the standard provided by touch, but rather that provided by vision, and we also seek to reduce what is provided by touch to the visual standard" (op. cit., p. 71). Wherever the reciprocal evocation of tactual and visual elements leads to a particularly noticeable conflict, e.g., in the size estimation of three-dimensional forms, one tends to speak of sensory illusions.[13] The statements of Subject Kretzer cited above (Section 20) indicate the grotesque overestimation of tactual magnitude that can arise as a result of inadequate specific experience. I do not intend to furnish new evidence here for laws already well known and established in many places, laws involving the relationship of vision to touch. Rather, we turn, following these general introductory remarks, to the problem pointed out at the beginning of this section.

2. *Observations on blind and brain-damaged subjects.* The problem posed cannot be decided with the sighted (for when using these persons, how can one absolutely exclude all visual vestiges?), but perhaps might be with the congenitally blind. Two observers were available to me who, according to their own credible statements, had been blind from birth. The two were

brothers, one being a simple musician, and the other a competent organist, both of moderate intelligence, but with very little capability for psychological observation. I first repeated the basic experiment with them. They could recognize the differences in roughness or smoothness, even if not quite as reliably as our trained experimental subjects. The absolute judgments given on the smoothness and roughness of the papers also were completely sensible. Thus, if even the congenitally blind possess them, are not the images of smoothness and roughness therefore rooted exclusively in touch? We would do well to defer the answer to this question until after we have received further information below. (Incidentally, the basic experiment, as well as all further tests with the blind, seemed to confirm our view that the full potential of the touch sense is revealed exclusively in movement; the fingers of the two blind subjects were eager to touch and it was hardly possible to keep them still on the tactual material during the pauses for reporting.)

In a second experimental series, the blind subjects were presented with the tactual materials of Section 29. Results: the tactual performance was essentially the same as, if somewhat weaker than, that of the sighted subjects. Insofar as the recognition of these materials is based on temperature impressions, as described above, and the temperature sense presumably was normal in the blind subjects, the performance cannot be surprising. However, the sensible judgments obtained on hardness and softness cannot be attributed to the temperature sense. The blind subjects also seemed to possess an image of elasticity, as mediated by a rubber band or a steel spring.

In interpreting the results, we must recall Goldstein and Gelb's (op. cit., p. 4) stricture that one be very cautious in drawing conclusions about the underlying psychic processes from the actual *performance* of a blind subject. The notion of performance, which these two authors base on action, such as the modeling of the blind, no doubt can be extended here to the making of a judgment, for the judgment would, after all, determine what the blind person would have to do. This means we could not conclude without further ado from similar performance by the blind that smoothness, roughness, etc., are provided to them in the same way as to the sighted. What happens on this question when we consider the statements on surface properties by a patient of Goldstein and Gelb's who had a complete loss of visual representations? The pressure sensations of this brain-damaged patient were essentially intact, except for a general reduction. What could the patient still do, and what could he no longer do? With his eyes shut, the crudest localization of contact was no longer possible, the stereognostic ability was completely eliminated, and even the two-point threshold no longer was obtainable, since the duality of contact was no longer recognized no matter how far apart the two stimuli were set. According to Goldstein and Gelb, when using a moving touch organ, the patient nevertheless correctly recognized velvet, sponge, rubber, metal, and cotton as such, designated cellulose not completely incorrectly as fabric, leather, or paper, used the designations rough and smooth, and hard and soft in a plainly correct fashion, and could even indicate what was flexible. What else could this mean but that even with the complete loss of visual representations, the ability to report on the structure of the surfaces, with their identifying characteristics (*Spezifikationen*) and qualities (*Modifikationen*), remains intact? Thus, the tactual experiences treated here do not rely on vision for their structure, but are completely independent tactual entities, and therefore are also available to the person born blind. Goldstein and Gelb (op. cit., p. 72) have asserted "that the tactual impressions of a healthy person must be tinged qualitatively differently by the normally concurrent visual input. Thus, the copious visual images and fringes aroused in the sighted undoubtedly modify the tactual impressions so that they are phenomenally quite different from the corresponding representations of the congenitally blind or the patients of Goldstein and Gelb. But the visual effect is only secondary. When we think we can see the roughness or smoothness and the hardness or softness of a surface, or the brittleness or elasticity of an object, it is really the experience of touch that has brought us to this juncture, with the visual indicators only finding their expression through touch.[14] Thus, in our view, a person provided only with vision would not attain the representations treated here, but would have to be content with ascertaining whether a surface is glossy or dull, reflective or non-reflective, etc.[15]

According to Goldstein and Gelb, qualities provided by the sense of touch do not have spatial properties as such (op. cit., p. 73). Based on experiments he performed on Ahlmann, Wittmann emphatically agreed with this view, even holding that movement does not create perceived space. He asserted that "the blind person's own movement . . . is not given perceptually or even pictorially as a change in spatial position, but only dynamically as a vague sequence of changing qualitative impressions (joint, muscle, tendon, touch, resistance, warmth, and coldness sensations, etc.)." "Thus, what one could call a qualitatively ordained type of tactual space is not created for the blind even by movements."[16] The congenitally blind, according to this, can never come to have a *surface touch* in the strict sense of the term (we intentionally avoided using this term above); for such persons, the properties of surface structure develop only in an *historical* sequence. Ahlmann's and Wittmann's view about the "historical" construction of the tactual world of the congenitally blind agrees quite well with our theory about the participation of vibration sensations in the tactual phenomena treated here. After all, we referred emphatically to the particular temporal mode of appearance of the vibration sensations, which transform the persisting, visually provided properties of surface structure into an historically developing sequence. Goldstein and Gelb left open the question as to whether the impressions of hard and soft are exclusively tactual qualities, or are influenced by visual images of the spatial deformation of the skin. I rather think that the tactual and kinesthetic qualities, with their special temporal organization, determine these impressions, as well as the impression of elasticity. Certainly, the corresponding impressions of the sighted, being permeated by visual elements, are not phenomenally the same, but the tactual-kinesthetic quality yet determines their basic thrust.

Goldstein and Gelb's patient completely lacked the impression of movement. Now, if for him, as with the sighted, the surface structure is brought to life by objective movement of the touch organ, but the movement as such *cannot be perceived* (with the sighted, it is usually not noticed, or need not be), then that would confirm my thesis on the significance of objective movement as one of the purely *causal* factors in the genesis of impressions like smooth and rough, which provide us with surfaces.

What about the modes of appearance of tactual impressions that I previously distinguished in Sections 5 to 8? Are they also discriminated by our two blind subjects? According to their statements, which were not very clear, this appears to be the case. If, like Steinberg, we were to adopt Heller's basic view "that the decisive factor for the adequate apprehension of three-dimensional forms is tactual movement, whose phenomenal correlate, however, only acquires a purely extensive organization when it is related to a three-dimensional image, as that alone gives simultaneous touching," a view that completely predominated until the publications by Goldstein and Gelb, then we would not hesitate to accept the possibility of a purely tactual genesis of the modes of appearance mentioned. However, we take the position here of Goldstein and Gelb. Therefore, in the construction of these modes of appearance we regard everything specifically spatial as having been imparted by vision. This naturally holds, too, for the structure of thermal impressions. However, given these results, can we still speak of modes of appearance of tactual impressions in the proper sense, and can we align them in parallel to the modes of appearance of color? Wittmann answered this question negatively (op. cit., p. 436); I emphatically answer it affirmatively. However, there is an inescapable need to describe the representations of the naturally developed forms or gestalten of tactual perception found in the consciousness of the normal sighted individual. If one is not biased theoretically, then the tactual gestalten we distinguished present themselves with the same primitiveness as the modes of appearance of color. The theory should and will receive its proper due eventually, but it seems a questionable procedure to set it up at the outset without considering the array of psychological facts favoring it from a purely phenomenological viewpoint. One might well ask whether precisely such a procedure has not often severely retarded the development of psychology. The atomism in sensory psychology that we have fought against in this book is also a result of such a procedure. Nevertheless, there is no external, readily recognizable, and easily decipherable association between tactual impressions and those of vision that lends spatiality to the tactual impressions. Rather, the permeation is such that only an unusual case of brain damage, like that found by Goldstein and Gelb, could shake the general consensus that tactual impressions have something spatial in them initially. In 1844, Hagen had indeed presented such a theory, but it remained simply a forgotten theory, which Goldstein and Gelb had to remind us of again.

I cannot readily agree, however, with Wittmann's (op. cit., p. 438) assertion hereto that "even the sighted can be convinced by tests on themselves of the reality of these non-spatial, purely qualitative contact and tension experiences, whose differences fade with prolonged motionlessness, and the fact that one can turn his attention energetically to his actual perception only after his eyes have long been shut; one will recognize that perception per se provides no point of departure at all for any sort of spatial representation." For example, when I spread my hand out on the table and observe its state for a relatively long period, then the clarity of the impression does indeed decrease considerably as fatigue sets in, but I cannot say that the spatiality of the tactual impression, its being spread out, disappears completely.

I should like to call attention to how the splitting off demanded here of all spatial quality accruing to touch through vision would affect research. All experiments hitherto conducted on the two-point threshold would then no longer be included in touch, indeed not even experiments on the simple pressure threshold. After all, Goldstein and Gelb (op. cit., p. 27) showed how even the value of the pressure threshold decreases *many-fold* under the influence of the visual image. The dependence of this supposedly tactually-based magnitude upon vision reveals like nothing else the typically indissoluble fusion of touch with vision. Thus, the tactual impressions do not permit the spatial garb that they have seized upon to be taken from them so easily. From all this, I see no alternative for research, even given the theory of Goldstein and Gelb, than to start from tactual forms with their spatial properties. Perhaps one day we will meet a brain-damaged person who is indeed still able to sense colors, but not with the spatial qualities that constitute their normal modes of appearance. Should it then be forbidden, from that moment on, to use the modes of appearance of color that I have distinguished as a starting point for research?

I have contended that the normal sighted person is unable to denude the tactual surface impression of its spatial properties simply by a change in mental set. That contention no longer holds, however, for surfaces that are touched at a distance using the probe principle. If I touch an object with a rod, then there is indeed an unanalyzable surface impression at the place of contact between hand and rod, but not at that between rod and object. The latter surface impression is subject to analysis when the naive mental set is dispelled, however. If I behave analytically, then its spatial character dissipates, leaving behind kinesthetic and vibratory sensations. With the normal sighted person, therefore, it is only directly by the pressure sense, [not indirectly by the probe,] that the visual fringes can be excited so markedly that the resultant spatial properties of the tactual impression become tough and resistant. Thus, in all of the research reported in this paper on touching at a distance, the vibration sensations indicating roughness and smoothness are referred to a merely image-like tactual surface.

Nothing visual corresponds to the vibration experience that indicates the *state* of an object. The movement must be so rapid to produce a vibration sensation that it is no longer perceived visually as such, but rather, as already mentioned, can at best be inferred from the blurring of the contours of the moving object. The attempt to visually image a state of vibration does not work; where it appears to succeed, closer examination shows that internally one has slowed the movement considerably.

The exposition in this and the following section documents the considerable influence of individual experience on the structure of the world of touch, but at the same time alludes here and there to the limits of this influence. Special circumstances exist in this respect in every area of sensation, and without a precise knowledge of them, the general study of the "nativism versus empiricism" issue must remain unproductive.

Footnotes

1. W. Steinberg, op. cit., p. 144.
2. E. Bonaventura, Gesichts- und Tastsinn in der Raumwahrnehmung (Vision and touch in spatial perception), *Riv. di psicol.*, *17*, 1921.
3. A. B. Fitt, Groessenauffassung durch das Auge und den ruhenden Tastsinn (Size perception by eye and motionless touch), *Arch. f. d. ges. Psychol.*, *32*, 1914.
4. In many of his papers cited.
5. E. R. Jaensch, Ueber Taeuschungen des Tastsinns (Touch illusions), *Zeitschr. f. Psychol.*, *41*, 1906.
6. H. Rupp, Ueber Lokalisation von Druckreizen der Haende bei verschiedenen Lagen der letzteren (Localization of pressure stimuli at different positions on the hands), *Zeitschr. f. Sinnesphysiol.*, *41*, 1907.
7. C. Spearman, Die Normaltaeuschungen in der Lagewahrnehmung (Normal illusions in position perception), *Psychol. Stud.*, *1*, 1906.
8. T. Ziehen, Ueber die Abhaengigkeit der scheinbaren Groesse taktiler Empfindungen von der Entfernung und der optischen Einstellung (Effect of distance and visual set on the apparent size of tactile sensations), *Zeitschr. f. Sinnesphysiol.*, *50*, 1916.
9. E. Gellhorn, Untersuchungen zur Physiologie der raeumlichen Tastempfindungen unter Beruecksichtigung der Beziehungen des Tastraums zum Seh-

raum (Physiology of spatial tactual sensations considering the relationships of tactual space to visual space), *Pfluegers Archiv*, *189*, 1921.
10. W. Ahlmann, Zur Analysis des optischen Vorstellungslebens (Analysis of visual imagery), *Arch. f. d. ges. Psychol.*, *46*, 1924.
11. A. Sophie Rogers, Auditory and tactual perceptions: The role of the image, *American Journal of Psychology*, *34*, 1923. In studies on dreams, A. J. Cubberley showed how easily tactual excitation (stimulation of a spot on the skin through an adhesive-tape intermediary) during a dream can evoke a characteristic visual situation. The effects of tensions of the body surface upon the normal dream, *British Journal of Psychology*, *13*, 1923.
12. Lillien J. Martin, Ueber aesthetische Synaesthesie (Aesthetic synesthesia), *Zeitschr. f. Psychol.*, *53*, 1909. We can call briefly to mind the innumerable investigations on the role of the tactual factor in the enjoyment of sculpture. "See with the feeling eye, feel with the seeing hand" (Goethe, *Roman elegies*).
13. O. Klemm, *Sinnestaeuschungen* (Sensory illusions), Leipzig, 1919. Klemm does not include in his list the generally little-known illusion that when one cuts into some soft wood with a knife under a strong magnifying glass, the resulting visual enlargement gives rise to the impression that one is cutting deeply into a soft mass, such as cork.
14. "Not only do the shape and form of the objects we see contain much that depends on the sense of touch, but we also *see* many properties now that actually were initially perceived only by the sense of touch and only through touch have become known: the brittleness and flexibility, the weight and heft of a material, and more." Haas, op. cit., p. 64.
15. The experience of so-called basket cases, i.e., persons born without all four limbs, can naturally not be used here, since they possess the touch sense on other portions of their body. For more on such persons, see G. Hirth, *Energetische Epigenesis* (Energetic epigenesis), Munich and Leipzig, 1898.
16. J. Wittmann, Ueber Raum, Zeit und Wirklichkeit (Space, time, and reality), *Arch. f. d. ges. Psychol.*, *47*, 1924, p. 436 f.

Section 47. The Influence of Kinesthetic Processes

I have already occasionally touched upon the question of the influence of kinesthetic factors on the phenomena described. It is time now to examine the question once again with a more systematic, even if brief, presentation, because it is related in part to the examination just completed of the contribution of visual processes to the fashioning of tactual forms. Based on their experiences with their patients, Goldstein and Gelb proposed that even in the tactual space of the normal person, visual images are not directly aroused by excitation of the sensory areas of the skin, but depend on the effect of kinesthetic fringes. If so, then the question of the role of kinesthesis for tactual forms has thereby been largely answered. The kinesthetic factor would affect tactual forms insofar, at least, as it mobilizes the visual fringes, which in turn participate in the fashioning of the tactual forms. Undoubtedly, however, the kinesthetic contribution is not limited to this indirect effect.

We have not tired of pointing time and again to movement as a creative force for tactual forms. Even if, as many contend, we are not conscious of the excitation of motor relay stations in the cerebral cortex, the effect of movement on the touch organs nevertheless does reach consciousness through the kinesthetic sensations.[1] The kinesthetic sensations, insofar as they become conscious, contribute to the tactual forms; they participate when the elements of the rough-smooth and soft-hard dimensions arise during active touching, as well as when a judgment on stickiness, oiliness, etc., is based on the experience of the resistance overcome during movement. We have already emphasized the exceptional independence of all these impressions from the force of movement used; this indicates that the intensity of the kinesthetic impression in the total experience was taken very precisely into account. With roughness and smoothness, hardness and softness, etc., the muscle-sense component of kinesthetic sensation stays wholly in the background. It cannot be completely disregarded, but we know from cases of passive touching that it can be omitted completely, with scarcely any detriment to the preciseness (*Praegnanz*) of those impressions. The relationship between the two components (muscle, tactual) of the kinesthetic sensation reverses itself, however, with the elasticity experience, whose clarity rises and falls with the clarity of the muscle-sense sensation, whereas the tactual component completely recedes, indeed, can even be completely omitted. Tapping experiments revealed to me a considerable sensitivity for the discrimination of degrees of elasticity. This is not surprising, considering that von Frey found that the muscle sense (sense of effort) surpasses all other senses with respect to the precision of discrimination.[2]

Bechterew[3] reports on a very remarkable "illusion" of the sense of touch with rotary vertigo: Rough surfaces appear smooth to the touch. Thus, the sense of balance also participates in the fashioning of tactual phenomena, along with the senses already discussed.

Footnotes

1. The sensations of the muscle sense are "those sensations evoked during muscle activity by the firing of sensory receptors that terminate in the muscles, tendons, etc. These sensations serve primarily to inform us on the degree of tension in our muscles (the so-called sense of effort). By contrast, perception of the position and movement of our extremities is based largely on the cutaneous touch sensations. The so-called kinesthetic sensations therefore include both muscle and tactual sensations." G. E. Mueller, *Abriss der Psychologie* (Outline of psychology), Goettingen, 1924.

2. M. von Frey, Die Vergleichung von Gewichten mit Hilfe des Kraftsinns (Weight comparison with the aid of the sense of effort), *Zeitschr. f. Biol.*, 65.
3. W. von Bechterew, *Arch. f. Anat. u. Physiol.*, 1896, p. 105.

Editor's Notes on Division IV: Applications (Sections 48 to 50)

In the next (and final) division, Katz discusses miscellaneous issues not covered in detail in previous sections. Some issues he discusses involve applications, such as on the use of the sense of touch in education (Section 49) and in psychological tests (Section 50). Katz was dissatisfied with the extent of training and testing done on touch, especially that involving the vibration sense and texture. How unnatural it is, he observed, for the school-age child to be "forced to keep his or her touch-thirsty fingers still in a well-behaved manner on the bench during the longest periods of instructions" (Section 49).

Other issues he discusses involve more theoretical matters, such as the subsections in Section 48 on linguistic forms and the primacy of touch. Katz invokes the classical philosophical arguments (Krueger, 1982) favoring the primacy of touch (it is the sense first used by the infant, the sense upon which physics is based, the sense used to dispel visual ambiguities or illusions, etc.), but without providing any fresh data of his own. More recent data indicate that vision generally dominates and educates touch, not vice versa (Rock & Harris, 1967). More convincing perhaps, is Katz's argument that the causal connection between object and percept is stronger for touch than vision. As a result, the causative nature of verbs of sensory activity in German, whose objects are in the accusative case, is based on touch, with its stronger interpenetration of subject and object or environment. Katz also provides a fascinating analysis and listing in Section 48 of cognitive words derived from the sense of touch (e.g., grasp, handle, comprehend). This is a good point. The historical prominence of touch is evident in English, too, in the way touch-related words have been extended to other modalities, such as in "sharp tastes," "dull sounds," and "soft colors." Rarely does the reverse occur; we never talk, say, of "loud or fragrant touches" (J.M. Williams, 1976).

DIVISION IV: APPLICATIONS

Section 48. The Psychology of Language and Perception

1. *Linguistic expressions derived from the tactile-motor sphere, particularly the activity of the touching hand.* When we wish to indicate that we have mastered some facts of a practical or theoretical nature, we say that we have *grasped* them. This word, like the noun, *grasp,* derived from it, and the adjective, *graspable,* points unmistakably to the manual activity by which sensory-motor mastery is achieved over an object (to grasp = to span with a grasp). An emotion or activity that I can see through and understand appears *graspable* to me. Along with *grasp,* there are words in German not directed exclusively to the mental realm, such as *Angriff* (attack), *Eingriff* (operation), *Zugriff* (grip), and the illness *Griff* (cold); in German these words, like *begriffen* (grasp), all derive from the root *Griff* (grasp), and thus appear to echo acoustically something of the forceful grasping of the fist in both its brevity and sharpness. We should also mention briefly here, too, the derivations *angreifbar* (vulnerable) and *griffig* (handy). A more general form of the mental tendency underlying *grasp* is expressed by the verb *erfassen* (comprehend), and next to it the vulgar *kapieren* (catch); both words again point to the activity of our grasping organ. If we survey the German language, we see that in the overwhelming majority of cases the images depicting a mental relationship between subject and object are borrowed from the tactile-motor sphere, particularly from the activity of the touching hand. The instrument of mental acquisition, which also produces grasping, is *Vernunft* (reason), a word whose derivation from manual activity is less obvious than that of the corresponding verb, *vernehmen* (perceive) [*nehmen* = to take]. A *faehiger* (capable person) is one who can *fassen* (seize) something. The verbs *zergliedern* (dissect), *behalten* (contain), and *auslegen* (display) refer both to actual manual activity and, in a figurative sense, to mental activities. *Ueberlegen* (consider) is used only with reference to mental acts. With *stellen* (place), which in German occurs as part of *vorstellen* (imagine), one can probably also think of a hand movement.

The words *handeln* (handle), *Handel* (trade), and *Handlung* (activity) are tied directly to the root *hand;* these designations reveal what attention the activity of the inborn universal instrument has found in practical life. The compound words *Abhandlung* (treatise) and *Verhandlung* (negotiation), which also are tied to the root *hand,* point more to the mental domain.[1]

Expressions derived more generally from the tactile-motor area are just as numerous as the linguistic images tied to manual activity that were cited above. I mention first: *it touches me,* then with increasing intensity, *it grasps me, it hits me, it does not let me go, it goes all the way through me.* I can *take something to heart.* The word *Eindruck* (impression), which indicates in the most comprehensive manner an action on our mind, is also borrowed from the

DIVISION IV: APPLICATIONS   239

tactile sphere [*Druck* = pressure]; its counterpart, *Ausdruck* (expression), likewise is tied to the tactile-motor domain.

As far as I can see, besides the (naturally, incomplete) list of verbal expressions derived from the sense of touch to indicate intellectual processes, only a few have been borrowed from other sensory domains. *Einsehen* (observe), *einbilden* (imagine), and *scheinen* or *erscheinen* (appear) point to the visual domain. What causes this predominance of images derived from the tactile-motor domain? It is probably based on the primacy of touch compared to all other senses in perceptual psychology. Before furnishing evidence for this, examples from other languages will show that the verbal expressions examined are not simply a peculiarity of the German language.

In Latin, the verbs *percipere* (perceive) and *comprehendere* (understand) refer to understanding by means of the hand. *Oblivisci* (forget) likewise is connected through its derivation from *oblinere* (obliterate) to a manual activity, just as is *fingere* (*animo sibi* or imagine oneself), whose original meaning was to knead. *Cogitare* (think) is derived from *agere* (act, the act of). Does not the contrast between *videri* (see), with its visually-derived activity, and *manifestus* (palpable, obvious), provide a prime example of the utmost clarity for the present thesis? The verbs *imprimere* and *exprimere* correspond completely in meaning to the German verbs *eindruecken* (impress), *einpraegen* (impress), and *ausdruecken* (express).

Recall that in Greek, *katalepsis*, with its derivation from *lambanein*, corresponds to grasp, take possession of.[2] Derived from the same root are *analambanein* and *npolambanein* (comprehend, grasp, understand mentally). The word "character" is derived from *charasso* (sharpen, scratch). Mere appearance or sensory illusion is designated by *opsis*, which is taken from the sense of vision.

Linguistic images derived from manual activity or the tactual domain in general are frequent in French and English, owing to the predominance of Latin roots in these modern languages. French: *comprendre, concevoir, capable, exprimer, imprimer*. English: comprehend, conception, capable, impression, expression. In concluding the consideration of the derivation of linguistic images from the tactual domain, let me merely add that the present exposition appears to be valid for all European languages. It would be a commendable exercise to do corresponding analyses for languages from other linguistic families as well.[3]

2. *The primacy of touch in perceptual theory*. In the Preface, I called into question the usual division of the senses into higher and lower levels in psychology textbooks. The sum total of the investigations presented here, I would hope, provide a basis for my position, at least insofar as touch has been numbered among the lower senses. We are not dealing here with a pointless semantic exercise, nor with an internal concern of sensory psychology. No,

the division has had, and still has, particular import for perceptual theory. General questions of the psychology of perception ought not to be studied solely from the perspective of vision and audition, disregarding what touch reveals in its particular fashion, and the one-sided orientation of perceptual theory towards the higher senses is just as regrettable from a philosophical viewpoint. Henning has quite properly spoken out against the dismissal of the sense of smell as a lower sense in psychology and philosophy, just as we do here with respect to the sense of touch.[4]

The sense of touch (in which, for simplicity of exposition, we also include kinesthesis) indeed does not provide all of the subtle nuances available in vision. Also, remote sensitivity, which has reached full development in vision, is found in only rudimentary form in touch. Even so, from a perceptual viewpoint we must give precedence to touch over all other senses because its perceptions have the most compelling character of reality. Touch plays a far greater role than do the other senses in the development of belief in the reality of the external world. Nothing convinces us as much of the world's existence, as well as the reality of our own body, as the (often painful) collisions that occur between the body and its environment. What has been touched is the true *reality*[5] that leads to *perception*; no reality pertains to the mirrored image, the mirage that applies itself to the eye. The rod held immersed in water appears broken to the eye; the hand corrects the error of the eye. Despite variation in retinal size and shape, and variation in apparent surface structure, an object has only one true size and shape, and only one true surface structure. This fact is indicated by the tactual impression, which does not vary when the hand enclosing the object gives it a different distance from the eye. No doubt, space would not be structured with reference to the up-down direction in the way actually observed without the cooperation of touch. Based on such considerations, Locke and Berkeley gave touch precedence over vision, and based visual space on tactual space. "This sense is also the only sense of direct external perception, and precisely for that reason it also is the most important sense and the one that can be most reliably taught" (Kant). According to Schopenhauer, too, in the final analysis the visual percepts refer to those of touch.[6] Many philosophers who did not explicitly express such thoughts nevertheless can be shown to have included them in the background for their theories of perception.

We already mentioned previously that such basic concepts of physics as impenetrability, resistance, and force are rooted in touch, to which we now add the concept of friction. Physics, and with it natural philosophy, would not have taken their present historical form if we had not been equipped with touch. The physics of persons deprived of touch would probably be very different than ours; the physics of the blind and of the deaf do not differ from ours.[7] Distinctive, too, is the physics that the believers in ghosts have developed for ghosts. Ghosts can be seen and heard, but have had the heavy and impenetrable body taken from them. It is precisely the body that, like nothing else, makes us conscious of our ties to physical reality.

Human development also indicates the primacy of touch. For the infant at the threshold of life, the pressure sensations, which greatly intensify during birth, can most readily provide the experience of reality of all of the sensory impressions storming in. Then, from the first day on, there are tactual experiences of sucking, which occur in connection with the vital process of taking in food. "The *original space* of the newborn is the region of the *mouth*; the mouth is presumably the only organ that, from the very first day on, responds to definite tactual impressions with definite movements."[8] Even without active touching, tactual impressions constantly occur in the infant, owing to the pressure of clothing, the ground, and the body care received. They occur to such an extent that all other sensory impressions, in their scope and intensity, must be overshadowed. In addition, the temperature impression so important for the behavior and well-being of the newborn, must be included here in the tactual impression.

As soon as the child learns to use its hands, a true passion for touching awakens. An important improvement in the touching tool occurs with the opposition of the thumb, which Preyer observed in his child at the age of 3 months. Our own two children started to touch and grasp all the objects that the environment offered them at about 11 months. It is no exaggeration to say that whatever the child sees, it will also want to touch.[9] The child relies on the tactual impression, which alone seems to guarantee the reality of the object. Preyer's child was puzzled by the absence of tactual feeling with a mirrored object.[10] Sikorski's child saw that the biscuit box was empty, but grabbed into it anyway in order to convince himself that it contained nothing. Just as with visual hallucinations, touch provided a check on reality. This tendency even persists in the behavior of adults.

3. *The accusative case for verbs of sensory perception.* Jaensch devoted a section of his book on the world of perception and its structure in youth,[11] to the relationship between eidetic imagery and philology. He attempted to explain why verbs of perception occur in connection with the accusative case [in German and other languages], inasmuch as perception is neither a mere activity nor an action, and nothing is performed or produced in perceiving.[12] The question had previously been raised by C. Sigwart in his work on logic. Rejecting other possible solutions, Jaensch reached the conclusion "that no involved assumptions, indeed generally no hypothesis at all, is necessary, even with the verbs of perception, in order to point out a causal relationship in the primitive consciousness, as the causative demands." "With eidetic imagers, the perceptual function actually exercises quite literally a causal role vis-á-vis the objects of perception, by influencing, changing, indeed even constructing them." Jaensch refers to his finding that the eidetic imager feels directly engaged during visual perception, and

that the objects are altered as a result of the visual attention. Jaensch uses ethnological material as further support for his view, in that human development generally seems to reveal an eidetic phase in visual perception. "As to the objection that all of these relationships primarily hold only for visual perception, we would merely point out, first, that in all primitive languages and early stages of development, visual percepts definitely dominate the verbally-expressed contents of perception . . . and, second, that the corresponding relationships have not been explored in more detail yet in the other senses, which is not to say that they will not be found there." In psychopathology, Storch believes he has confirmed Jaensch's position with schizophrenic patients. "The original percept has a dynamic character; perception deals with the given sensory facts in the sense of a causal function."[13] "With schizophrenic patients . . . we can observe how the rigid subject-object contrast in the experience of the patient disintegrates, and is replaced by a magical cause-and-effect relationship. Perception for such patients is not a simple glance at an object standing opposite the ego, but rather an acting out of the ego upon the sensorially given, or even an acting in upon the ego of the sensorially given."

I believe that an analysis of the tactual experience can help solve this problem. The circumstances of tactual perception reach more deeply than do those of visual perception towards the roots of the linguistic expressions discussed here. It has not been firmly determined that we pass through an eidetic phase of visual perception, and therefore it remains only an hypothesis that the linguistically creative nature of precisely this period manifests itself in the accusative construction of the verbs of perception. It is much more natural to tie these linguistic constructions to tactual rather than visual perception, because no assumption of an eidetic phase is required here at all. Every ongoing tactual activity represents a production, a creation in the true sense of the word. When we touch, we move our sensory areas voluntarily, we *must* move them, as we are constantly reminded, if the tactual properties of the objects are to remain available to us. Who would keep the hand still while touching in everyday life? Such behavior is demanded only by the instructions of the laboratory. The tactual properties of our surroundings do not chatter at us like their colors; they remain mute until we make them speak. By our muscular activities, we *produce*, so to speak, such properties as roughness and smoothness, and hardness and softness. We are truly the *creators of these qualities*, and not only the micromorphic [substance] qualities [e.g., roughness, hardness], but also, to a large extent, the macromorphic [shape] ones, as the data on stereognosis show. We point out once again here how the circumstances of the tactual image reflect this fact very clearly. Touching means to bring to life a particular class of physical properties through our own activity. One could therefore deduce almost a priori that only the causative linguistic form can do justice to the psychological structure of the tactual process. If we start with the activity of touching, then we need not seek the aid of a (still hypothetical) eidetic phase of perception to explain the psycholinguistic phenomena. (Furthermore, Jaensch may be correct in stating that even the sense of touch once passed through an eidetic phase.) The creative nature of tactual perception is a fact that can be confirmed by anyone at any time through introspection.[14]

It is not only the activity and causation of the tactual process that make it appear better suited than visual perception to explain the linguistic forms being discussed; support also is obtained from an additional result of our earlier analysis of tactual perception. Storch considers the cases of pathological experience he describes as particularly conclusive evidence for Jaensch's position, because in these cases the chasm between subject and object that ordinarily exists in perception has not yet opened up. Where the object is not set off as an invariant opposite the ego, but rather the ego acts in the environment without sharply differentiating its own efforts, the linguistic expression that depicts the world as a gushing forth of the creative activity of the ego actually appears quite apt. Thus, for Storch, the schizophrenic's form of perception represents a reversion to an earlier stage of phylogenetic development, whose linguistic form for the activity of perception strikes us as inadequate. Many things speak in favor of Storch's hypothesis, which is based on experiences of perception that, unlike normal perception, permit no sharp distinction between subject and object. It provides further justification for us to point again to tactual perception as exhibiting a decided bipolarity. With touch, no hypothesizing at all of earlier stages of development is needed; even in the consciousness of normal human beings, tactual perception never occurs without the subjective pole. Storch presumably thought primarily of visual perception, where the dissociation from experience actually goes so deep that, as already mentioned, Hering even called the term visual *sensation* inappropriate. Normal visual perception is not interweaved with what is experienced. It is different with the sense of touch. Recall our detailed consideration of the bipolarity of the tactual impression, in which the touch organ cannot be excluded. Object and subject cannot be imagined at all as separate factors of the tactual impression. If one sees in that type of interpenetration of environment and subject the prerequisite for the development of the linguistic causative form, then it is, in any case, given much more compellingly in tactual experience than visual perception.

Granted that what has been said up to now could explain the causative linguistic form for the verbs that refer to the activity of touching, what about the corresponding linguistic expressions for the other areas of perception? Jaensch posited transference from vision to other sensory domains, owing to the preeminence of vision. I think that it is more plausible psychologically for the spreading to occur from touch. The preceding paragraph dealt with the primacy of touch from very different viewpoints. We have attempted to demonstrate that touch is often called upon in the final analysis to decide the matter when visual verification cannot be achieved or illusions arise, and that this tendency, which appears instinctively even in children, is pursued self-consciously by adults. Therefore, we posit transference of linguistic forms developed in the tactual domain to other sensory domains, on account of the primacy of touch within the whole province of perception.

Footnotes

1. The word *finger*, which is connected to the activity of the finger, has played a large role, especially in the jargon of soldiers. [So-called] primitive persons and children *finger* numerical representations. *Verstehen* (understanding) and *Verstaendnis* (comprehension) took reference from the activity of the lower extremities [*stehen* = to stand].
2. "Chrysippus . . . .and along with him Antipater from Tarsus and Apollodorus and others posit as a criterion of truth the *kataleptike fantasia*, i.e., that image which, proceeding from a real object, forces the assent of the subject and thereby produces a *katalepsis*." Ueberweg-Heinze, *Grundriss der Geschichte der Philosophie des Altertums* (Outline of the history of ancient philosophy), 9th ed., Berlin, 1903, p. 293.

   I cannot help quoting some remarks by Aristotle on the human hand. "Anaxagoras indeed asserts that it is his possession of hands that makes man the most intelligent of the animals; but surely the reasonable point of view is that it is because he is the most intelligent animal that he has got hands. Hands are an instrument; and Nature, like a sensible human being, always assigns an organ to the animal that can use it. . . . We should expect the most intelligent to be able to employ the greatest number of organs or instruments to good purpose; now the hand would appear to be not one single instrument but many, as it were an instrument that represents many instruments. . . . All the other animals have just one method of defence and cannot change it for another: they are forced to sleep and perform all their actions with their shoes on the whole time, as one might say; they can never take off this defensive equipment of theirs, nor can they change their weapon, whatever it may be. For man, on the other hand, many means of defence are available, and he can change them at any time, and above all he can choose what weapon he will have and where. Take the hand: this is as good as a talon, or a claw, or a horn, or again, a spear or a sword, or any other weapon or tool: it can be all of these, because it can seize and hold them all. And Nature has admirably contrived the actual shape of the hand so as to fit in with this arrangement." (Aristotle, *Parts of animals*, IV, 10; translated by A. L. Peck. Cambridge, Mass.: Harvard University Press, 1968, pp. 371-373.)
3. A few other relationships between speech and manual activity will be pointed out here. The motor centers for the hand and speech lie immediately adjacent to each other in the brain. The speech center for right-handed people, who use

mainly the right hand to gesticulate during speech as well as to touch (and also to produce tactual images), is in the left brain, where the right hand has its main representation. In this connection, we can also mention the remarkable finding by Saer that bilingual children in Wales who spoke both Welsh and English were three times more likely to encounter difficulty with the distinction between left and right than were monolingual children. D. J. Saer, The effect of bilingualism on intelligence, *British Journal of Psychology*, *14*, 1923.

4. H. Henning, Assoziationsgesetz und Geruchsgedaechtnis (Association principle and odor memory), *Zeitschr. f. Psychol.*, *89*, p. 192. H. Henning, *Der Geruch* (Odor), 2d ed., Leipzig, 1924.
5. Henning remarks in his study on the psychology of thinking cited above (*Zeitschr. f. Psychol.*, *81*, 1919, p. 78): "All reported tactual images in all series concerned ... something *real*. It should also be pointed out that even the tactual fantasies in fiction, e.g., the tactual games of Martians (Kurd Lasswitz, *Auf zwei Planeten* [On two planets]), likewise refer back to something real or at least possible."
6. A. Schopenhauer, *Ueber den Satz vom Grunde* (Principle of sufficient reason), Section 21.
7. W. Lay believes that a person who completely lacks kinesthetic impressions cannot be successfully reared. *Zeitschr. f. exp. Paedagog.*, *3*, 1906.
8. W. Stern, Die Entwicklung der Raumwahrnehmung in der ersten Kindheit (Development of space perception in early childhood), *Zeitschr. f. angew. Psychol.*, *2*, 1909, p. 413.
9. "At the age of three months, the child commences to use his hand for grasping; he palpates like a new connoisseur, and the tendency to test, to search for tactile-muscular sensations develops in him day by day." Perez, *Les trois premiéres années de l'enfant* (The child's first three years), 5th ed., 1892, p. 38 f.
10. W. Preyer, *Die Seele des Kindes* (The soul of the child), 7th ed., 1908, p. 353.
11. E. R. Jaensch, *Ueber den Aufbau der Wahrnehmungswelt und ihre Struktur im Jugendalter* (The world of perception and its structure in youth), 9. Abschnitt (Section 9).
12. This question must be restricted to those languages, like German, in which verbs of activity prevail almost entirely. It makes no sense for languages like Georgian, which distinguish quite sharply between verbs of activity and verbs of sensation. On this, see F. N. Finck, *Die Haupttypen des Sprachbaus* (The main types of linguistic structure), Leipzig, 1910.
13. A. Storch, *Erlebnisanalyse und Sprachwissenschaft* (Experiential analysis and philology), *Zeitschr. f. Psychol.*, *94*, 1924, p. 147 f.
14. A certain activity unfolds as a rule when we execute the eyemovements that serve visual perception, but these eyemovements do not create color the way finger movements create touch.

Section 49. The Sense of Touch in Education

Since the cultivation of the senses has become part of the educational curriculum, attention to training of the sense of touch has indeed not been lacking, but as a rule it still has been treated quite like a stepchild compared with the so-called higher senses. The extensive passage quoted from *Émile* above (Section 39, Footnote 5), where Rousseau suggests that the sense of vibration be exercised, provides an exception to the rule that only one branch of the sense of touch, that of pressure, was considered in the training of "feeling," as it was usually called in popular psychology. Rousseau makes numerous suggestions on the training of the sense of pressure and its stereognostic capabilities. Chiefly, form recognition was presented as a task for touch. Thus, in the older pedagogy of the normal child, an exercise such as classifying plants exclusively by the feel of their leaves (as Salzmann[1] had done on occasion in the Philanthropin zu Schnepfenthal), whereby the surface structure also had an important part to play, was an utter curiosity. The so-called Froebel gifts for the kindergarten speak to the sense of form of the eye and of the touching hand; the type of surface of the wooden blocks is completely secondary.

The systematic training of the senses receives a wider scope in the very important Montessori[2] system of education than in any of its predecessors. She provides detailed directions and exercises for stimulating the sense of touch. The child is offered a rich collection of tactual materials in order to acquaint it with the most varied surface structures. It is easy to get children interested in the exercises, and Montessori properly points out the true passion of the young child to let all available objects pass through its examining hand. Montessori has the exercises carried out with the eyes shut. Because the ears are left open, this is not strictly sufficient to ensure a purely tactual performance, but pedagogical hairsplitting is not appropriate here; the sense of touch undoubtedly can be trained in this way as well. We can only underscore what this fine pedagogue says about the importance of training the hand for various occupations.

The evidence that movement actually provides the vital element for the sense of touch has important pedagogical implications. It means that the school-age child is quite unnaturally and harmfully constrained when forced to keep his or her touch-thirsty fingers still in a well-behaved manner on the bench during the longest periods of instruction, instead of allowing them freedom of movement to acquire new, systematically-directed, tactile-motor experiences.

With the blind, and even more so with blind deaf-mutes, the training of the sense of touch takes on a very particular significance; this is well enough known so that it is sufficient here merely to mention it.

Progress in the psychology of perception has generally had a stimulating effect on pedagogics, so I hope that the research presented here will be of some use to all of those concerned with training the sense of touch, whatever their viewpoint. Where in everyday life could one not make use of a well-trained sense of touch? It certainly is no tragedy, as experienced housewives have assured me, when woollen things occasionally are patched with cotton (and vice versa), something that could not be done if the sense of touch were better trained. It is worse, though, when, due to insufficient training, the hand of the physician fails during palpation and in other instances. The cultivation of the sense of touch in sighted persons is always limited, in reality somewhat arbitrarily, to the hand. Why should the other touch organs not also be exercised? Persons born without arms (the leg artist Untan), and those who have lost hands by amputation, show us by their dexterity what tactual feats the feet and stumps can perform when harsh necessity so dictates. Touching in everyday life is always touching in motion; its perfection is at least as much a motor as a sensory matter. On the other hand, there is scarcely any exercise for dexterity that, while emphasizing motor activity, does not also exercise sensory touching.[3] A formal goal of vocational instruction is to increase manual dexterity. I believe that many findings in this book can contribute to the unfinished theoretical foundations of vocational instruction. The lack of coherent pedagogical principles in *psychological testing* is revealed more strongly in those portions covering tests of the tactual ability of the hand.

Footnotes

1. *Salzmanns Ameisenbuechlein* (Salzmann's pocketbook of ants), Verlag F. Schoeningh, Paderborn, 1912, p. 51.
2. Maria Montessori, *Selbsttaetige Erziehung im fruehen Kindesalter* (Spontaneous activity in education during early childhood), Verlag J. Hoffman, Stuttgart, 4000-7000, p. 173 f.
3. Hitschmann (op. cit., p. 392) pointed out how the heightened perfection of the sense of touch in the blind usually results in a corresponding increase in manual skills. Thus, blind persons learn quite rapidly how to reproduce various fruits and leaves in clay or wax. W. Wirth also pointed out significant sensorimotor capabilities of the human hand in his psychophysical analysis of the registration of the Repsold micrometer. *Wundts Psychol. Studien, 10,* 1917.

Section 50. The Sense of Touch in Psychological Testing

Psychological tests in which the tactual aspect of performance predominates, are treated in Rupp's[1] brief survey of aptitude tests as demonstrations of the "sensitivity of the hand," but all of the tests on dexterity, which he puts after them, likewise contain a tactual component. In addition, Rupp does not mention the test for sensitivity to differences in texture, which requires that a set of sandpapers of varying fineness or metal surfaces of varying granularity be put into order. This test has obviously done well. Skutsch reports, based on examination results for the national railroad, "that failures in the sense of touch, found in 1921 in the comparison of sandpapers of varying granularity, were markedly below the national average, a sign of the occupational significance of this property, to which attention will be paid."[2] From my own theoretical viewpoint, performance on these tests is determined wholly or in part by vibration sensitivity. Compared with the best performance obtained in our basic experiment, that attained or so much as requested on the aptitude tests using series of sandpapers would be called only moderate. The basic experiment used papers having very small differences, which made the test maximally difficult; for this reason, it could assist in setting up a new and more rigorous standard of the performance to be required in this kind of test. In addition to this, the numerous variations of the basic experiment, as well as the research on the identifying characteristics (*Spezifikationen*) of the tactual impression, will perhaps stimulate the applied psychologist to devise new tests.[3]

We come now to the test of the ability to recognize flatness and unevenness. A difference in level must be recognized at the sharp edge between two flat surfaces. W. Moede built a special apparatus, the so-called sense-of-touch examiner (*Tastsinnpruefer*), for this task. Less well known is that this device had a forerunner in an apparatus described by Graham Brown[4] in 1902. O. Lipmann and Stolzenberg arranged the levels in very finely gradated steps, the difference in height being .02 mm. In a much different connection, O. Klemm called attention to the fact that when two smooth pieces of metal are jammed against each other to form a straight line, the probing finger that is moved to and fro over this line can even recognize differences in height between the two surfaces of less than .02 mm.[5] As surprisingly small as the difference in level is that is normally recognized during this test, it considerably exceeds in size the irregularities in the plane that determine the impression of roughness or smoothness of many surfaces, as shown by our findings above.

Another test of the sensitivity of the hand requires arranging a number of cardboards, metal sheets, or wires according to thickness. Since the test materials never went below a certain thickness, the findings cannot reveal that astonishing sensitivity for very thin material reported above (Section 31). Our relevant results can perhaps also find use in psychological testing.

The specimens that Rupp lists under No. 31 (the precise recognition of the friction of a bolt or a threaded bolt, or of the pressure of a threaded bolt

DIVISION IV: APPLICATIONS 249

against a turning stroke) and No. 32 (precise recognition of the friction of a blank) of his set, appear to be directed at the fineness of resistance sensations. From my own experience, however, they also involve sensitivity to vibrations transmitted through a rigid intermediary. With the test that requires the ordering of springs of varying elasticity, we touch on the questions mentioned in Section 47; those tests that require tactual recognition of spatial forms lie beyond the scope of the present work.

Footnotes

1. H. Rupp, Eignungspruefungen (Aptitude tests). Section 12, *Taschenbuches fuer Betriebsingenieure* (Pocketbook for operating engineers), Berlin, 1924.
2. Skutsch, Die psychotechnische Versuchsstelle der Reichsbahn (The psychological testing office of the national railroad), *Prakt. Psychol.*, 1923, p. 325.
3. The manager of a sugar factory sent me the following very noteworthy letter on the use of the hand in cooking sugarbeets. "When the cooking apparatus has been filled, then the finished brew must be drawn off at a certain concentration. This can be determined by the cook only by feeling. He takes a sample from the cooking apparatus and rubs it between the fingers. He then discerns the viscosity and thereby the concentration, and with such precision that the difference between his test and a laboratory test is usually only 0.5% (e.g., instead of 94%, it is about 94.5% dry substance). Making these tests naturally requires much practice, and not every cook can learn it. For this reason, these sugar cooks are invariably salaried employees and well paid."
4. L. Tigerstedt, *Handbuch der physiologischen Methodik* (Handbook of physiological methodology), *III, 1*, p. 18. Leipzig, 1914.
5. O. Klemm, *Sinnestaeuschungen* (Sensory illusions). Leipzig, 1919, p. 39.

Author Index

Abderhalden, E., *39*
Achelis, J.D., *43*
Ahlmann, W., 226, 228, *232*
Allers, R., *129*
Alrutz, S., *39*
Ammann, J.C., *192*
Arisotle, *244*
Arnheim, R., 1, 12, *17*
Aubert, H., *52*

Baade, W., *37*
Baker, A.S., *82*
Ballieu, 188
Balss, H., *208*
Baltzer, F., *208*
Bard, L., *222*
Barker, L.F., *177*
Barrovecchio, 188
Basler, A., *108*, *197*
Bechterew, W. von, *233*
Békésy, G. von, 2, 6, *17*
Benussi, V., *71*, *87*, 88
Berkeley, 240
Bernstein, 197

Billigheimer, *175*
Bing, *188*
Blix, 34
Boas, *123*
Boehm, *67*
Bonaventura, E., 226, *231*
Boring, E.C., 6, 13–14, *17*
Braus, H., *36*
Brentano, F., 12, 70
Brown, G., 248
Brunswik, E., 14, *17*
Buehler, E., 40, *42*
Buerklen, K., *67*, 116, 120, 220

Capraro, A.J., 8, *17*
Cerulli, 188
Churchill, 226
Cobbey, L.W., *39*
Cohn, L., *150*, 221
Collins, A., *20*, 224
Cooper, C., 11, *18*
Cornelius, H., *76*, *77*, 83
Cubberley, A.J., *232*

Dallenbach, K.M., 6, *21*, 184
Darian-Smith, I., 8, *20–21*, 184
Darlington, C., 15, *21*, 223
Darwin, 195
Déjerine, 188
Deschamps, *192*
Dichtl, A., *207*
Dodge, R., *86*
Donaldson, 34

Ebbecke, U., *43*, 59, *61*, *177*, *180*, 181
Ebbinghaus, H., 38, *39*
Edinger, L., *208*
Edmonds, E.M., *220*
Egger, *188*
Elliott, J., 15, *21*, 223
Elze, C., *36*
Erdmann, B., *86*
Erismann, T., *71*
Eschke, E.A., 189, *193*, 220
Ettlinger. M., *196*, *205*, *211*
Ewald, R., 197, 209, *211*
Exner, S., 86, *87*

Feldt, J., 192, *194*
Felix, *212*
Féré, C., 28, *30*
Fick, 180
Finck, F.N., *245*
Fitt, A.B., 226, *231*
Forli, 188
Foulke, E., 2, *21*
Frank, C., 188, *192*
Frey, M. von, 6–7, 12, *21*, *36*, 39, 62,
 77–78, *79*, *82*, *99*, *120*, *173*, *180*,
 188, *192*, *197*, *199*, 200, *202*, *203*,
 210–211, *212*, 226, *233*
Friedlaender, H., 42, *43*
Frisch, von, 206, *208*
Froebel, 246
Fuchs, W., *54*, *61*
Funke, O., *36*

Gelb, A., 49, 226–227, 230, 232
Geldard, F.A., 2, 6–7, *17–18*, 223
Gellhorn, E., 226, *231*
Gescheider, G.A., 5, 8, *18*

Gibson, J.J., 1–6, 8, 13–15, *18*, 31, 73
Giese, F., 29, *30*
Gildemeister, 81, *82*
Goethe, 232
Goldmann, A., 78, *79*
Goldscheider, 34, *82*, 122, 150, *180*,
 188, *202*, 220
Goldstein, K., 39, 49, 159, *160*,
 226–227, 230, 232
Goodwin, A., 8, *20*
Gordon, G., 2, 7, *18*
Gordon, I.E., 11, *18*
Grandis, V., 197
Green, B.G., 9, *18*, *99*, 164
Groos, K., 108
Grote, L.R., *36*
Guttmann, W., *187*
Gutzmann, H., 191, *193*, 195

Haas, W., *203*, 232
Hacker, F., *100*
Haecker, 207, *208*
Hahn, J.F., 6, *18*
Halpern, F., *100*, *129*
Hansen, K., *194*
Happich, K., *120*
Hardenberg, K. von, *120*
Harris, C.S., 2–3, *21*, 235
Hase, A., *220*
Hauptmann, G., *30*
Hausmann, T., 121, *123*, *150*, 214
Head, *192*
Heidegger, M., 12
Heller, M.A., 2–4, 8, 10, *18*, 32, 140
Heller, S., *79*, 90, 149, 229
Helmholtz, H., 75, 77, *87*, 209, 210
Henning, H., *67–68*, 149, 240, *245*
Henri, V., *119*, 226
Hering, E., 34, *42*, 67, 84, 150, 176–177,
 243
Hermann, 197
Herzog, 188
Hill, A.V., *194*
Hirth, G., *232*
Hitschmann, F., *221*, *247*
Hoefer, P., 82, 122, 202, 220
Hoffa, A., *123*

AUTHOR INDEX

Hoffmann, H., *119, 141*
Hoffmann, P., *194*
Hofmann, H., *36*
Holm, 174
Hornbostel, E.M. von, *83, 88*, 194, 205, 212
Husserl, E., 12, 70

Jaensch, E.R., *37*, 83, *84*, 226, *231*, 241–243, *245*
James, W., 15, *18*, 184
Javal, E., *99*
Jennings, J.L., 12, *18*
Jesionek, A., *211*
Johansson, R.S., 7, *21*
Jones, B., 3, 8, *20*, 32
Judd, 77, 226

Kafka, G., *208*
Kammler, O., *52*
Kant, E., 5, 28, 240
Katz, D., 1, 3–12, 14–17, *18–19*, 23, 27, 29, *30*, *42*, 51, 54, 67, 85, 98, 110, 115, *119*, 123, 138, 140, 159, 187
Katz, R., *30*
Keller, H., 108, 189–190, *193*
Kennedy, J.M., 2–3, 11, 16, *19*
Keyserling, H. von, *120*
Kiel, 189
Kiesow, F., *52*, *120*, *180*, 200, *203*, *209*, 211
Klatzky, R.L., 2–3, 5, 8, *19–20*, 45, 224
Klemm, O., *208*, *232*, 248, *249*
Koehler, O., *206*, 210
Koehler, W., 28, *30*, 192, *194*
Koffka, K., *37*, *87*, 206, *208*
Kornhuber, H.H., 8, *21*, 184
Kramer, 188
Kries, J. von, *180*
Kries, L. von, *173*
Krogius, A., *180*
Kroh, O., *68*
Krueger, L.E., 1, 3, 14, *19*, 62, 235
Kuelpe, O., 42

Lamb, G., 8, *19*
Landolt-Boernstein, *173*

Lasswitz, K., *108, 245*
Laubi, O., *194*
Lay, W., *245*
Lederman, S.J., 2–3, 5–11, *18–21*, *32*, 45, 62, 90, *99*, 129, 144, 164, 184, 224
Lindner, R., 192, *194, 206*
Lipmann, O., 248
Locke, J., 225, 240
Loehner, L., *175*
Loomis, J.M., 2, 6, 7, 10, *20*
Lotze, H., 28, *30*, 81, 114, 121, *123, 203*

Mach, E., *67*, 84–85, 108, 195, *196*
Mack, A., 14, *20*
MacLeod, R.B., 1, 4, 12–13, *19–20*
Malmud, R.S., *180*
Marinesco, 188
Martin, L.J., 226, *232*
Martius, G., 29
Matthes, E., *87, 207, 208*
Meinong, A., *70*, 77
Meissner, 203
Metzner, R., 77, *79*
Michotte, A., 14, *20*
Mingazzini, 188
Minor, 188
Moede, W., 29, *30*, 248
Montessori, M., 246, *247*
Morely, J., 8, *20*
Mountcastle, V.B., 8, *21*, 184
Muck, O., *196*
Mueller, G.E., *67*, 77, *233*
Mueller, J., 210

Nagel, W., *39*
Natsoulas, T., 14, *20*

Oppenheim, 188

Parish, 226
Pauli, R., *62*
Pereire, J.R., 192
Perez, *245*
Perky, C.W., 42
Peters, W., 77
Petzoldt, J., *52*

Pfingsten, 189
Pillsbury, 226
Plate, E., *200*
Plessner, H., *203*
Ponzo, M., *180*
Popper, E., 159, *160*
Preyer, W., 241, *245*

Quilliam, T.A., 7, *20*

Rabaud, 206
Raphel, G., *192*
Redlich, 188
Reed, C., 3, *19*
Reid, T., 8, 13, *20*
Révész, G., 2, *20*, 220
Rock, I., 2–3, 14, *20–21*, 235
Rogers, A.S., *232*
Rousseau, *192*, 246
Rubin, E., *58*, 149
Rumpf, 188
Rupp, H., 42, *43*, 78, 226, *231*, 247–248, *249*
Rydel, 188

Saer, *245*
Sahli, 122, *123*
Salzmann, 246, *247*
Schaefer, K.L., *61*
Schaer, A., *194*
Schapp, W., 42, *43*, 108
Scheibner, 29, *30*
Schiff, W., 2, *21*
Schiller, 24
Schlesinger, G., *30*
Scholl, K., *192*
Schopenhauer, A., 240, *245*
Schottelius, E. von, *173*
Schottmueller, *123*
Schumann, F., *208*
Schumann, P., *192*
Schwaner, *192*, 197
Seiffer, 188
Sergi, G., *199*
Sherrington, C.S., *36*, 177
Shults, E., *82*
Sigwart, C., 241

Sikorski, 241
Sklar, B.F., 8, *18*
Skramlik, E. von, *180*
Skutsch, 248, *249*
Smith, M.E., *220*
Spalteholz, *178*
Spearman, C., 226, *231*
Steinberg, W., 160, *220*, 226, 229, *231*
Stephenson, W., 16, *19*
Sterling, 188
Stern, W., 77, 83, 86, *87*, 189, *193*, *245*
Stevens, J.C., 9, *18*, *99*, 164
Stoerring, G., *69*, *199*
Stolzenberg, 248
Storch, A., 242–243, *245*
Straub, E.R., *84*
Stumpf, *194*
Sullivan, A.H., *39*
Sutermeister, E., 189–191, 200

Talbot, W.H., 8, *21*, 183
Taylor, M.M., 9, *21*, 90, 129
ten Horn, C., *30*
Thalman, A. W., *85*
Thompson, S., 86
Thorne, G., 3, 8, *20*, 32
Thunberg, T., 38–39, *174*, 176
Tigerstedt, L., *39*, 188, *197*, *249*
Titchener, E.B., 35, *36*, 37, 45, 47–49, 76
Treitel, 187–188
Truschel, 220

Ueberweg-Heinze, *244*
Untan, 140, 247

Valentin, 188
Vallbo, A.B., 7, *21*
Van Doren, C.L., 8, *18*
Verrillo, R.T., 7–8, *17–18*, 21
Vierordt, K., 188, *192*
Voigt, A., *178*

Walker, J.T., 6, *21*
Wardell, J., *20*, 224
Washburn, 226
Webels, W., *212*

AUTHOR INDEX

Weber, E.H., 33, *36*, 41–42, 66, 75, 100, 107, 126, 143, 148, 177, 195, 226
Weiskrantz, L., 15, *21*, 223
Weizsaecker, V. von, *212*
Wertheimer, M., *83*
Williams, D.A., 6, 10, *20*
Williams, J.M., *21*, 236
Williams, J.T., *119*, 189
Wirth, W., *247*
Wittich, von, 188
Wittmann, J., *68*, 228–230, *232*
Wood, J.E., *197*
Worchel, P., 6, *21*, 184
Wulff, O., 83, *84*
Wundt, W., 35, 209, *211*

Zech, F., 221
Ziehen, T., *205*, 226, *231*
Zigler, M.J., 1, *21*, *120*
Zwislocki, J.J., 8, *17*

Subject Index

Adaptation, 6, 78, 99, 111, 127–129, 174, 199
After-images, 199
Amputees, 64, 67–68, 121, 138–140, 171, 216
    Hallucinatory images (phantom limb), 68, 159
Animal psychology, 86–87, 206–208, 220
Appearance, modes of, 15–16, 47–49
    Tactual, *see* Immersed touch, Surface touch, Volume touch
    Visual, *see* Color (Film, Surface, Volume)
Apsychology, 70
Astereognosis (tactual object blindness), 10, 140, 227
Audition, 83, 195, 200, 221
    Auditory cues in touch, 11, 114, 125–126, 187, 214
    vs. touch, 6, 24, 191–192, 195–196, 204–206

Bipolarity of tactual phenomena, 14, 40–41, 62, 243

Blind, 40, 67, 78–79, 82, 99, 116–117, 119, 121, 149, 150, 160, 179, 201, 206, 220–221, 225–227, 246
Body as tactual object, 126–127, 168, 198, 200, 214, 244

Color,
    Film, 50–52, 98
    Surface, 50–52
    Volume, 52–53
Constancy, perceived, *see also* Invariant impressions of object or surface,
    Pitch, 220
    Position, 84–85
    Properties of object, 84–85
    Roughness, 9, 216–217
    Temperature, 11–12
Continuity of tactual surface, 59–61, 157–160

Deaf, 6, 187, 189–194, 200–201, 206, 209, 220

Distal (remote) sense, touch as, *see also* Proximal (near) sense, touch as, 5, 25, 117, 176, 184, 204–208
Doppler effect, 217–218

Echolocation, 6, 184
Elasticity, judgment of, 16, 51–52, 73, 80–82, 114, 233
Episcotister, 110–111

Figure and ground, 60–61
Film, touch-transparent, 53–54, 145
Foot as touch organ, 124–125, 140, 190–191, 200
Form, *see* Macrostructure

Gestalt psychology, 1, 8–9, 12, 42, 126, 171–173, 175, 210, 229

Hand as touch organ, 2–5, 28, 45, 63–64, 109, 126, 134, 138, 145, 159, 190–191, 244, 249
Hardness, judgment of, 3, 13, 15, 80, 100, 113–114
Hearing impaired, *see* Deaf
Heat box, 168–171

Identifying characteristics (Spezifikationen) of surface touch, *see* Surface touch
Images, tactual, *see* Memory touch
Immanent vs. transeunt observations, 134–135
Immersed touch, 15–16, 50–52, 164, 180
Intermediaries, tactual, 10–11, 53–54, 61, 109–115, 118, 121, 128, 140, 148–149, 215, 219, 230
Invariant impressions of object or surface, *see also* Constancy, perceived, 9, 11, 14–15, 73, 80, 198, 233

Kinesthesis, 35, 80–81, 87, 204–205, 223, 229, 232–233, 240
Kinesthetic cues in touch, 223, 232–233

Language, 43, 55, 58, 203, 235–236, 238–239, 241–244
Lips as touch organ, 123–124, 168, 199

Macrostructure, 3, 8, 40, 86, 126, 242
Memory touch, 62–68, 101
Methodological issues, 31–35, 77–78, 89–90, 93, 130–131, 145, 153, 160
Microstructure, 3, 8–9, 40, 54–58, 85–86, 242
  vs. color, 8, 54, 57–58
Modalities of touch, *see* Kinesthesis, Pain sense, Pressure sense, Spatial sense of skin, Temperature sense, Vestibular sense, Vibration sense
Molineux problem, 225
Monotony of touch matter, 48
Movement, 5, 15, 37–39, 63, 65, 73–88, 95–96, 100–101, 106, 108, 128, 151, 153–161, 227, 233, 246
  Active vs. passive, 105–107
  Finger, 151, 156

Objective pole, *see also* Projection, external; Subjective pole, 13, 16, 41–43, 68, 79, 99, 164
  Movement, 15, 41
  Tactual, 51, 68, 99, 126–127
  Visual, 13–14

Pacinian corpuscle, *see* Vibration sense
Pain sense, 35, 37, 42–43, 82, 196
Palpation, *see also* Volume touch, 53–54, 121–123, 149–150, 220
Percussion method, 17
Perspiration and touch, 11, 107, 174–175
Phantom limb, *see* Amputees
Phenomenal aspects of touch, 12–17, 28, 39–42, 77, 202, 209–210, 229
Pressure sense, *see also* Spatial sense of skin, 5–12, 35, 37–38, 48–49, 70, 77, 81–82, 117, 121, 160, 177, 179, 189, 197, 200, 203, 209, 241
  Adaptation, 78, 199
  Intensity, 49, 137

SUBJECT INDEX

Projection, external, *see also* Objective pole, 13, 114, 121
Proximal (near) sense, touch as, *see also* Distal (remote) sense, touch as, 5, 40, 203–204

Qualities (Modifikationen) of surface touch, *see* Surface touch

Reaction time, 131–134, 170, 173, 198, 201
Recognition of tactual material, 130–141, 165–171, 213
Reduction procedures, *see also* Adaptation, 9–10, 89–90, 98–99, 138
  First-degree, 98–99
  Second-degree, 98–99
  Size, 96–97
  Suppression of lateral movement, 100–101
Roughness, judgment of, 3, 9–11, 15, 80, 99–100, 106–109, 111–112, 115–119, 164, 214–215, 218

Sensory psychology,
  Atomistic viewpoint, 9, 33–35, 66–67, 76
  Tachistoscopic mentality, 4, 76
Shape, *see* Macrostructure
Skin, 33, 48, 62, 75, 82, 116, 139, 173, 197, 209, 219
Smoothness, *see* Roughness, judgment of
Softness, *see* Hardness, judgment of
Sound, *see* Audition
Spatial perception, 83, 119
Spatial sense of skin, *see also* Pressure sense, 9, 35, 116–119, 122, 150, 217, 219, 230
Specific nerve energy, Mueller's Principle of, 7, 211–212
State vs. property of object, 119, 164, 179–180, 184, 199, 212, 219

Stickiness, judgment of, 16, 118–119
Subjective pole, *see also* Objective pole, 13, 41–43, 68, 79, 99, 164, 178
  Tactual, 41, 51, 68, 99, 126–127
  Visual, 13, 41
Surface touch, 50, 55, 69, 93, 164
  Identifying characteristics (Spezifikationen), 46, 55–56, 69, 79, 89, 98, 100, 106–107, 130–141, 228, 248
  Qualities (Modifikationen), 46, 55, 69, 79, 89, 93–129, 228

Talbot's law, 197
Telepathic recognition, 116, 120, 126
Temperature gestalt of tactual material, 171–172, 175–177
Temperature sense, 11–12, 35, 37, 42, 59, 128, 136, 163–181, 241
Temporal perception, 83, 201, 213, 228
Texture, *see* Microstructure
Theory of objects (Gegenstandstheorie), 69–70
Thermal Invariance Principle, 167–169
Thickness, judgment of, *see also* Volume touch, 3, 145–149
Toe as touch organ, *see* Foot as touch organ
Tongue, 7, 150, 198, 214
Touch blends, 37–39

Vestibular sense, 205, 223, 233
Vibration sense, 5–12, 115, 117–119, 121, 126, 129, 183–222, 228, 230, 248
  Adaptation, 6, 78, 129, 199
  Localization of vibratory source, 5–6, 207
  Pacinian corpuscle, 7–8, 183
  vs. pressure sense, 6–7, 9–12, 183–184, 197–199, 209
Vision
  Animals, 86–87
  Movement, 84–86
  Retina, 59, 61, 64, 74, 84, 86, 183

Vision *(cont.)*
　Visual cues in touch, 2–3, 27, 39–42, 65, 133, 135–137, 223–232
　vs. touch, 2–3, 14, 24, 46–47, 50, 57, 60, 84, 110, 195, 197–198, 202, 225, 229, 230, 235, 240, 243

Volume touch, *see also* Palpation, 17, 52–53, 150, 164, 178–179

Weber's Law, 148
Weight, judgment of, 16–17, 180–181

BF 275 .K3713 1989
Katz, David, 1884-1953.
The world of touch

CANISIUS COLLEGE LIBRARY
BUFFALO, N. Y.

DEMCO